Stone Decay
Its Causes and Controls

EDITED BY

Bernard J. Smith   and   Alice V. Turkington
School of Geography   *Department of Geography,*
Queens University of Belfast   *University of Kentucky*

Proceedings of Weathering 2000
An international symposium held in Belfast
26–30 June 2000

LONDON AND NEW YORK

First published 2004 by Donhead Publishing Ltd

Published 2015 by Routledge
2 Park Square, Milton Park, Abingdon, Oxon OX14 4RN
711 Third Avenue, New York, NY 10017, USA

Routledge is an imprint of the Taylor & Francis Group, an informa business

© Taylor & Francis 2004

All rights reserved. No part of this book may be reprinted or reproduced or utilised in any form or by any electronic, mechanical, or other means, now known or hereafter invented, including photocopying and recording, or in any information storage or retrieval system, without permission in writing from the publishers.

Product or corporate names may be trademarks or registered trademarks, and are used only for identification and explanation without intent to infringe.

ISBN 13: 978-1-873394-57-1 (hbk)

A CIP catalogue record is available for this book from the British Library

Library of Congress Cataloguing in Publication Data has been requested for this book

Contents

	Acknowledgements	v
	Foreword	vii
	Contributors	ix
1	Introduction: the need for interdisciplinary thinking in stone decay and conservation studies *B.J. Smith and A.V. Turkington*	1
2	Quantification of the decay rates of cleaned and soiled building sandstones *M.E. Young, J. Ball and R.A. Laing*	13
3	Durability and rock properties *R.J. Inkpen, D. Petley and W. Murphy*	33
4	The use of image analysis for quantitative monitoring of stone alteration *V. Lebrun, C. Toussaint and E. Pirard*	53
5	Mechanisms of attack on limestone by NO_2 and SO_2 *G.C. Allen, A. El-Turki, K.R. Hallam, E.E. Coulson and R.A. Stowell*	75
6	Weathering of sandstone sculptures on Charles Bridge, Prague: influence of previous restoration *R. Přikryl, J. Svobodová and D. Hradil*	89

7   Experimental weathering of rhyolite tuff building stones   109
    and the effect of an organic polymer conserving agent
    Á. Török, M. Gálos and K. Kocsányi-Kopecskó

8   Chemical composition of precipitation in Kraków: its   129
    role in salt weathering of stone building materials
    W. Wilczyńska-Michalik

9   Interpreting the spatial complexity of decay features on   149
    a sandstone wall: St Matthew's Church, Belfast
    A.V. Turkington and B.J. Smith

10 Arkose 'Brownstone' tombstone weathering in the   167
    Northeastern USA
    T.C. Meierding

11 Weathering of Portuguese megaliths – evidence for   199
    rapid adjustment to new environmental conditions
    G.A. Pope and V.C. Miranda

12 Influence of anthropogenic factors on weathering of the   225
    Carpathian flysch sandstones
    W. Wilczyńska-Michalik and M. Michalik

13 Observations on stepkarren formed on limestone,   247
    gypsum and halite terrains
    D. Mottershead and G. Lucas

14 The role of mechanical and biotic processes in solution   273
    flute development
    D. Mottershead and G. Lucas

15 Weathering scales, landscapes and change: some   293
    thoughts on links
    R.J. Inkpen

Acknowledgements

The authors are indebted to the Queens University of Belfast Publications Fund for a grant towards the publication costs of these proceedings. Extensive cartographic assistance from Gill Alexander and editorial assistance from Patricia Warke, School of Geography, Queens University, enabled the production of this volume. The editors are grateful to all the referees who ensured the academic quality of the papers included here. Finally, thanks are due to Donhead Publishing for their long-term support of stone decay and conservation research.

Foreword

This volume presents selected papers from the Stone Weathering and Atmospheric Pollution NETwork conference, which formed part of the international symposium Weathering 2000, held in Queens University of Belfast, Northern Ireland, 26–30 June, 2000. This collection of papers demonstrates the breadth of research pertaining to stone decay conducted within disciplines including geomorphology, geology, materials science, chemistry and engineering. It also highlights the importance of communication between these disciplines and the value of an interdisciplinary approach to stone decay research.

The introductory chapter elaborates on the importance of intellectual exchange between researchers and practitioners. The following papers address the causes of, and controls on, stone decay in a range of environments. This is demonstrated by papers that investigate numerous aspects of stone decay and conservation, including the nature and rate of decay in both urban and rural environments. Several papers specifically address mechanisms of decay and synergisms between them; others deal with relationships between decay processes and decay forms. New methods of analyzing and quantifying decay are also described, and many authors have commented on the controls on decay exerted by stone properties and environmental conditions. Several stone treatments have been assessed under experimental conditions and the problems associated with inappropriate conservation strategies are documented. The volume also includes two papers that explore the conceptual background to studies of stone decay and durability. Most significantly, they emphasize the importance of individual research teams in setting research agendas and, through experimental design, influencing the results obtained and the uses to which they are put. In doing so they exemplify cross-disciplinary similarities in the research process and show how thinking outside ones discipline not only provides new insights but can also avoid the repetition of mistakes already encountered elsewhere.

We hope therefore, that this volume will prove informative to both researchers interested in stone decay and conservation and to practitioners. For both theoretical and practical reasons, successful conservation strategies can only be developed with a sound knowledge of the causes of stone decay and their controlling factors.

Bernard Smith
School of Geography, Queen's University of Belfast, UK.

Alice Turkington
Department of Geography, University of Kentucky, USA.

Contributors

G.C. ALLEN
University of Bristol
Interface Analysis Centre
Oldbury House
121, St. Michael's Hill
Bristol BS2 8BS
United Kingdom
E-mail: g.c.allen@bristol.ac.uk

R.J. INKPEN
Department of Geography
University of Portsmouth
Buckingham Building
Lion Terrace
Portsmouth PO1 3HE
United Kingdom
E-mail: Robert.inkpen@port.ac.uk

V. LEBRUN
Laboratoire MICA
Université de Liège
Avenue des tilleuls, 45
4000 LIEGE
Belgique
E-mail: vlebrun@ulg.ac.be
www.ulg.ac.be/mica/vl

T.C. MEIERDING
Department of Geography
University of Delaware
Newark
DE 19716, USA
E-mail: meierdng@udel.edu.

D. MOTTERSHEAD
Department of Geography
University of Portsmouth
Buckingham Building
Lion Terrace
Portsmouth PO1 3HE
United Kingdom

G.A. POPE
Department of Earth and Environmental Studies
Montclair State University
Upper Montclair
New Jersey 07043, USA
E-mail: popeg@mail.montclair.edu

R. PŘIKRYL
Institute of Geochemistry, Mineralogy and Mineral Resources
Faculty of Science, Charles University
Albertov 6
Prague 2, 128 43
Czech Republic
E-mail: prikryl@mail.natur.cuni.cz

B.J. SMITH
School of Geography
Queen's University Belfast
Belfast BT7 1NN
United Kingdom
E-mail: b.smith@qub.ac.uk

Á. TÖRÖK
Budapest University of Technology and Economics
Department of Construction Materials and Engineering Geology
H-1521 Budapest
Hungary
E-mail: torok@bigmac.eik.bme.hu

A.V. TURKINGTON
Department of Geography
University of Kentucky
1473 Patterson Office Tower
Lexington
KY 40506, USA
Email: alicet@uky.edu

W. WILCZYŃSKA-MICHALIK
Institute of Geography
Pedagogical University
30-084 Kraków, ul.
Podchorazych 2
Poland
E-mail: wmichali@wsp.krakow.pl

M.E. YOUNG
Masonry Conservation Research Group,
School of Construction, Property and Surveying,
The Robert Gordon University,
Garthdee Road,
Aberdeen AB10 7QB
United Kingdom.
E-mail: m.young@mailer.rgu.ac.uk

1 Introduction: the Need for Interdisciplinary Thinking in Stone Decay and Conservation Studies

B.J. SMITH and A.V. TURKINGTON

ABSTRACT

This brief introduction outlines the papers in this volume and sets them in the context of the aims of the Stone Weathering and Atmospheric Pollution Network (SWAPNET) conference at which they were presented. It identifies stone decay and conservation as a multidisciplinary field that requires interdisciplinary solutions. This is illustrated through the incorporation of concepts of episodic landscape change into explanations of non-uniform stone decay. It is further argued that successful conservation strategies require integrated consideration of societal as well as technical appropriateness. Finally, it proposes that sustainable solutions are best achieved if the contested ownership of heritage is recognized, and stakeholders are incorporated into the decision-making process through a partnership approach.

BACKGROUND

The origins of this book can be traced directly to the July 2000 meeting of the Stone Weathering and Atmospheric Pollution Network (SWAPNET) that took place under the auspices of the First International Weathering Conference held in Belfast. However, its lineage is more complex than this, and in reality it is the latest manifestation of a longer-running attempt to bring together two broad groups of researchers and practitioners with common interests in the ways that stones decay. Broadly speaking, these groups range from those interested in the weathering of rocks under natural environmental conditions through to those concerned with the decay and conservation of building stones – primarily within a heritage management context. The former group mainly comprises academic geomorphologists and geologists, whereas the latter is principally drawn from architects, engineers and conservators. Unfortunately, despite their obvious common interests, these groups generally lead parallel, separate existences with few opportunities for face-to-face contact and little in the way of intellectual exchange. This separation is manifested in many ways, not least at a terminological level, where those interested in the natural environment insist on referring to theirs as the study of 'rock weathering', whilst those with a concern for the built environment study refer to 'stone decay'. However, divisions run deeper than this and are maintained by attendance at separate conferences, publication in discipline-specific journals and membership of exclusive professional organizations.

INTERDISCIPLINARITY IN STONE DECAY RESEARCH

Although one can possibly dismiss the divisions within stone decay research as simply a failure to communicate, there are also deeper, philosophical divisions. One of the most fundamental of these is the largely academic pre-occupation with a theoretical search for explanations of why rock/stone decays, the processes and mechanisms responsible and the factors that control them. In contrast, conservation professionals are required to provide practical solutions through appropriate treatments and management strategies. Clearly, however, each approach needs the other. If academics fail to set process studies in the context of their wider, practical significance then they run the risk of pursuing the irrelevant. Similarly, a failure to accurately diagnose the causes of decay can often lead to conservators treating symptoms instead of causes. As Kant observed, application without theory is blind, whereas theory without application is mere intellectual play. In effect, this recognition of a need to link theory to application is a demand for co-operation between disciplines in weathering and conservation studies. Many see a multidisciplinary approach as the solution for this, but there is often confusion between the bringing together of multidisciplinary teams and true interdisciplinary research. Multidisciplinarity describes the physical make-up of the team and defines the nature of the problem. Interdisciplinarity hopefully describes the nature of the solution. It implies discourse between disciplines, synergies stemming from the integration of strategies, an appreciation by individual team members of where others are coming from and a willingness to borrow and learn new ideas. Without this discourse, a multidisciplinary approach may simply result in the accumulation of successive layers of mistakes contributed in turn by each discipline.

The value of interdisciplinarity can be demonstrated at many levels. At the philosophical level, paradigm shifts within disciplines are often externally driven. As Kuhn argued in his analysis of the nature of change in the sciences, for most of the time disciplines centre on paradigm reinforcement.[1] During these periods existing conceptual frameworks or dogmas are accepted and supported by their general application or by minor refinements that immunize them against criticism. Change from within the discipline is difficult as it is seen as a threat to the *status quo* and internal criticism may be stifled by, for example, a denial of access to peer-reviewed publications. The one ungoverned source for new ideas is, therefore, the adoption of innovations from other disciplines. Within the field of urban stone decay, the influence of other disciplines is exemplified by the way in which ideas on the nature and rate of decay over time have changed in recent years. If one goes back some fifteen to twenty years, it would have been quite common for rates of urban stone decay to be expressed in terms of extrapolated rates of long-term loss based on measurements on buildings over the two to three years of a typical research grant or on short-term laboratory experiments. This belief in decay as a gradual process that can be accurately predicted from limited observation, has its origins in the early uniformitarian view of natural geological processes in which the present was seen as the key to the past.[2] This in itself drew upon biological ideas of gradual evolution. However, within geological and geomorphological sciences, the interpretation of change in earth systems as being slow and incremental has come under much scrutiny in recent years, so that the behaviour of these systems is now seen as characterized by episodic change, in which long periods of quiescence are interrupted by short periods of rapid change as strength/stress thresholds are exceeded.[3] These thresholds can be breached through the gradual accumulation of internal stresses and/or a gradual reduction in strength. Alternatively, they may be triggered by a high magnitude external stress such as a major flood, or in the case of rock weathering, a severe frost. This episodic view of geological and geomorphological systems is analogous to the concept of 'punctuated equilibrium' in the biological sciences championed by researchers such as Stephen J. Gould[4] and for natural systems as a whole in the science of self-organized criticality propounded by Bak.[5]

Acceptance of episodic change as a model for system behaviour clearly makes prediction of future change problematic and over short- and medium-term scales it renders concepts such as average rates of change largely redundant. What is interesting is that threshold concepts and episodic change are also now finding their way into explanations of how stone decays (see Turkington and Smith, Chapter 9).[6] This is especially the case for non-calcareous stones such as quartz sandstones that may exhibit little overt sign of damage for many years before experiencing rapid, catastrophic loss through processes such as contour scaling (see Meierding, Chapter 10 for an example).

Even for calcareous stone, solution loss can be usefully thought of as an episodic occurrence triggered by rainfall. Acceptance of the episodicity of decay highlights the effects of individual rainfall events, which in turn focuses attention on the significance of parameters such as frequency, duration, intensity, variations in rainfall chemistry and the overall amount of rain as well as how these vary over time. In doing so it undermines simplistic assumptions of the averaging of solution rates based upon artificial constructs such as mean annual rainfall. Temporal variability in process rates can also be translated into a variable spatial response of stone to exposure (see Turkington and Smith, Chapter 9), which may be related to subtle differences in the weathering environment at the stone/atmosphere interface. Consequently, current understanding of the nature of spatial and temporal variations in decay has lead to a greater appreciation of the differing behaviour of stone within architecturally complex structures (see Young *et al.*, Chapter 2) and has stimulated interest in microclimatic conditions on, around and inside buildings. Often these conditions differ greatly from the generalized regional climatic data supplied by meteorologists.

The adoption of concepts of environmental variability and the non-uniform response of natural systems has therefore strongly influenced current thinking on the nature of stone decay. Most of all, however, it has focussed attention away from descriptions of how stone has decayed towards explanations of why decay occurs and what causes it.

INTERDISCIPLINARITY AND CO-OPERATION IN CONSERVATION STUDIES

The need for an interdisciplinary approach to stone conservation has been accentuated in recent years by recognition of the inherent complexity of stone decay, which often combines physical, biological and chemical processes under the control of geological and environmental factors.[7] This is even before any consideration of the cultural, societal and institutional contexts within which stone is used and which conservation strategies must be employed. Of course, complex decision-making is not unique to stone conservation and many formalized structures exist for the reconciliation of competing interests within a problem-solving framework. One approach that appears to be particularly applicable to the identification of conservation strategies, is that of so-called 'Appropriate Technology'.[8] This has been used extensively in, for example, the field of technology transfer to less developed countries in areas such as water supply and sanitation provision.[9] However, it can be applied in principle to many other situations where societies and organization are asked to absorb innovative practices or technologies. In essence, the approach seeks to identify the factors that control the acceptability of a strategy in any particular situation. It then establishes appropriateness criteria for each of these parameters against which the innovation can be assessed. The most obvious of these parameters is technical feasibility, i.e. whether the strategy is capable of doing the job. The approach then recognizes that, even though a strategy may be technically feasible, it will ultimately fail if it is inappropriate at, for example, a social, economic, political, cultural, educational or aesthetic level. The methodology therefore requires identification of the controls on acceptability within each of these dimensions, investigation of the issues attached to them and, most importantly, an analysis of how they interact. Only then is it possible to identify which strategy is appropriate or what has to be done to make a strategy appropriate. The virtue of this approach is that it neither prescribes nor proscribes any particular solution. Instead, it requires any investigator to keep an open mind as to the most appropriate way forward, including entertainment of the do nothing option.

The need to keep an open mind is particularly apparent within the field of stone conservation, where recent years have seen heated debate as to how conservation should be approached.[10] Unfortunately, this has tended to produce a polarization of views.

On the one hand there are those who would favour a 'soft engineering' approach, in which 'time is seen as an artist', and which proposes that stone should be allowed to reflect its post-emplacement history. This school of thought generally favours minimal, reversible intervention, prefers amelioration of the environmental causes of decay and requires careful monitoring of stone condition. To use the analogy of coastal protection, it could be argued that this group favours a form of 'managed retreat'. This approach is especially applicable to prestigious, culturally valuable structures that have access to sufficient funds to support targeted research, ongoing monitoring and careful maintenance. At the other extreme, there are those who view aggressive mechanical or chemical intervention as a valid form of conservation. They are content that successful intervention, involving such procedures as cleaning, dressing-back and consolidation is not only possible, but that these procedures should be used to return buildings as far as possible to their original design and finish. Underpinning this is recognition that stone is not immutable. This implies an acceptance that stones have a given lifespan, that the act of placing them in a building does not immunize them from decay and that replacement can be a valid form of conservation. This viewpoint is supported by the attitudes of many building owners who aspire to the complete removal of accumulated grime and the making good of any damage to the fabric. There is also the purely pragmatic consideration that large conservation industries exist in many countries and that they are continuously searching for employment. Again, it would appear that the resolution to this apparent conflict lies in the application of appropriateness criteria. These would recognize a range of possible solutions, the appropriateness of which depends upon an holistic analysis of the individual circumstances of the building or structure. Embedded in this approach is an understanding that the definition of what is an acceptable strategy can change over time. This may be a response to the development of improved treatments and procedures, but may also reflect changing cultural attitudes.

The need for appropriateness in conservation studies is highlighted by the iconic status of many buildings that are, or are perceived to be, in need of conservation. Even where individual buildings are perceived to have a lesser cultural significance, they may still form an essential component within a unique and valuable streetscape. In both instances, however, conservation is invariably complicated by contested ownership. At one level, there is the owner who has the title deeds in his, her or their possession, but frequently society as a whole will also lay claim to a broader 'ownership' of what it sees as part of its cultural heritage. To these must be added regulatory bodies responsible for enforcement of any protective legislation and countless others who might derive economic benefit from a building or site through, for example, heritage tourism. The disputed ownership of heritage is not, however, restricted to the built environment, and is perhaps better understood in the context of contested spaces and the management of protected landscapes such as National Parks.[11] Central to the successful management of these spaces is the identification of 'stakeholders' in the sites and the nature of their stakeholding. This information can then be used to establish a partnership approach to sustainable management that allows a variable degree of ownership to all interested parties. This can be manifested through involvement in the decision-making process, but also through individual participation in practical conservation.[12] This partnership approach only works where there is a shared vision for the site. But it is also more easily achieved where external financing is required to support conservation and where collaboration can be included as a requirement of the funding.

SWAPNET 2000

It is in the above spirit of intellectual exchange, the identification of appropriate remediation strategies and the forging of conservation partnerships, that the annual SWAPNET conferences have been held since 1995. The organization has its origins in the informal coming together in the late 1980s of a group of researchers, principally geomorphologists, who appreciated the similarities between weathering processes on natural rock outcrops, particularly in salt-rich arid and maritime environments, and those found in polluted urban environments. Important links are maintained with geomorphology, but attendance at the meetings has broadened to encompass not only a wide range of academic disciplines, but also researchers and stone conservation professionals. It is, however, gratifying that this distinction is becoming blurred, as more conservation projects come to appreciate the value of targeted research and academics increasingly seek to apply their research to actual buildings and monuments.

This 'integrated diversity' is reflected in the contributors and contributions to the Belfast meeting that constitute this volume. Of these contributions, there remains a core of studies into the impacts of pollution on stonework in urban environments. Three of these reflect a growing appreciation of the superb built heritage to be found in central Europe, but also of the damage inflicted on it during and after the Second World War. These studies come from the Czech Republic, Hungary and Poland (Chapters 6, 7 and 8), but the inclusion of a study of sandstone decay from Belfast (Chapter 9) serves to remind us that much of Western Europe also lives with a legacy of urban/industrial pollution. Studies of urban stone decay are complemented by two micro-scale analyses of natural limestone weathering by Mottershead and Lucas (Chapters 13 and 14) and of granitic megaliths by Pope and Miranda (Chapter 11), all from the Mediterranean region. However, there are also two studies of sandstone outcrops and buildings in the Carpathian Mountains (Wilczyńska and Michalik, Chapter 12) and sandstone headstones in the Northeastern USA (Meierding, Chapter 10) that examine the transition between urban and rural environments. The latter also provides interesting insights into the importance of micro-environmental controls on decay patterns. Methodological developments are not neglected and the papers by Young *et al.* (Chapter 2), Allen *et al.* (Chapter 5), and Lebrun, *et al.* (Chapter 4) deal respectively with the mapping of façades to elucidate rates of decay associated with conservation actions, the laboratory simulation of corrosion in polluted atmospheres and the quantification of colour change through a study of the artificial aging of building stone. The paper by Allen *et al.* (Chapter 5) also investigates the effectiveness of stone treatments within the laboratory, as does that of Török *et al.* (Chapter 7), which combines laboratory tests of a water repellent with a study of stone performance in use. The more philosophical aspects of scale issues in weathering studies and the durability concept are dealt with in the Chapters 3 and 15 from the University of Portsmouth, authored by Inkpen and colleagues. The papers presented at the SWAPNET 2000 meeting therefore ranged widely across the realms of rock weathering, stone decay and conservation. As such, they provided the participants, and now hopefully the readers of this volume, with numerous insights into the nature of weathering/decay processes and the factors that control them. Only when armed with this knowledge can we then go on to identify appropriate conservation strategies.

References

1. Kuhn, T.S. (1962) *The Structure of Scientific Revolutions*. University of Chicago Press, Chicago.
2. Thorn, C.E. (1988) *Introduction to Theoretical Geomorphology*. Unwin Hyman, London.
3. Schumm, S.A. (1979) Geomorphic Thresholds, the Concept and its Applications. *Transactions Institute of British Geographers*, 263: 10–119.
4. Gould, S.J. and Eldridge, N. (1977) Punctuated Equilibrium: the tempo and mode of evolution reconsidered. *Palaeobiology*, 3: 114.
5. Bak, P. (1997) *How Nature Works*. Oxford University Press, Oxford.
6. Smith, B.J. (1996) Scale problems in the interpretation of urban stone decay. In Smith, B.J. and Warke, P.A. (eds) *Processes of Urban Stone Decay*. Donhead Publishing, London; 3–18.
7. Viles, H.A., Camuffo, D., Fitz, S., Fitzner, B., Lindqvuist, O., Livingstone, R.A., Maravelaki, P-N.V., Sabbioni, C. and Warscheid, T. (1997) What is the state of knowledge of the mechanisms of deterioration and how good are our estimates of rates of deterioration? In Baer, N.S. and Snethlage, R. (eds) *Saving Our Cultural Heritage: The Conservation of Historic Stone Structures*. John Wiley and Sons, Chichester; 95–112.

8   Riedijk, W. (1987) *Appropriate Technology for Developing Countries*. Coronet Books, London.
9   Pacey, A. (1975) *Water for the Thousand Millions*. Pergamon Press, Oxford.
10 Maxwell, I. (1992) Stone cleaning – for better or worse? An overview. In Webster, R.G.M. (ed.) *Stone Cleaning and the Nature, Soiling and Decay Mechanisms of Stone*. Donhead Publishing, London; 3–49.
11 Goodman, D. and McCool, D. (1999) *Contested Landscape: The Politics of Wilderness in Utah and the West*. The University of Utah Press, Salt Lake City.
12 Riegert, M. and Turkington, A.V. (2003, in Press) Setting stone decay in a cultural context: conservation at the African cemetery No. 2, Lexington, Kentucky. *Building and Environment*.

2 Quantification of the Decay Rates of Cleaned and Soiled Building Sandstones

M.E. YOUNG, J. BALL and R.A. LAING

ABSTRACT

The rate of decay of building stone has an important impact on the long-term costs of façade maintenance. Factors that influence decay include stone characteristics, adjacent materials, pollution, climate and interventions, including stone cleaning. While the factors that cause stone decay are known, rates of decay are rarely quantified. By measuring the distribution of decay on approximately 150 sandstone façades, this study has quantified the relative rates of decay of cleaned and uncleaned building sandstones. Since this study was aimed at predicting future requirements of stone for repair it has concentrated on estimating the rate of stone decay in relation to progression in its surface coverage on façades. Results indicate that on uncleaned façades the mean rate of sandstone decay varied from 0.5 to 10.5% decay of surface area per century. On cleaned façades, decay rates were often accelerated, the response depending on sandstone type and cleaning method. For some sandstone types, decay rates could be increased by more than an order of magnitude. The mean rate for abrasively cleaned façades was approximately 42% surface decay per century; that for chemically cleaned façades approximately 61% surface decay per century. The response of individual façades could vary significantly from these means, depending on sandstone type, cleaning technique and the degree of care taken in cleaning. In a minority of cases stone decay rates may be reduced for a time following cleaning. The implications of cleaning in terms of its effect on rates of stone decay were found to be most prominent immediately after cleaning, declining over the following decade. These results show that cleaning can have a profound effect on the long-term behaviour of building sandstones and the potential long-term implications of cleaning should be fully appreciated before any decision is made to clean a building façade.

INTRODUCTION

Within the past few decades stone cleaning has had a significant effect on the appearance of building façades and it has recently become evident that stone replacement is being carried out on façades that have been previously cleaned. Cleaning is generally positively perceived by the public and building owners because of the simplistic notion that a clean, bright façade reflects well on the urban environment and on the image of the building occupier. However, no cleaning method has yet been devised which can remove soiling without affecting the underlying stone and the potential for immediate and longer-term damage to stone is seldom fully appreciated. Abrasive cleaning methods (e.g. grit blasting) inevitably result in some abrasive damage to the stone surface.[1,2] Chemical cleaning may dissolve some stone components and can leave substantial amounts of chemical residues in porous stone.[3] The soiling layer on building stone may be hydrophobic[4] and its removal is likely to change the behaviour of the stone (and joints) with respect to movement of water and water vapour.

QUANTIFICATION OF THE DECAY RATES 15

The amount of damage caused at the time of cleaning varies depending on cleaning technique and stone characteristics.[5,6,7] Cleaning also has possible consequences for the long-term behaviour of building stone as it has the potential to alter the rate of stone decay through physical and chemical changes to the characteristics of the stone surface. Although the immediate effects of cleaning have been documented[8,9,10,11,12,13] and attention has been drawn to the potential for further decay,[14,15,16,17] no previous studies have attempted to quantify the scale of this damage.

This research was commissioned by Historic Scotland to determine the extent and rate of stone decay on cleaned sandstone façades relative to uncleaned façades and to predict the effects of previous and on-going stone cleaning on the future costs of façade maintenance.[18] Objective and quantifiable evidence on the impact of stone cleaning with respect to stone decay and stone replacement is now urgently required if proper guidance is to be provided to both practitioners and the stone industry.

METHODOLOGY

A representative number of paired examples of cleaned and uncleaned sandstone buildings were selected from locations around Scotland (Table 2.1). Building pairs were normally adjacent properties whose façades were of the same age and were constructed with identical materials and design. Comparing the condition of paired building façades allows statistical analysis of the overall condition of cleaned and uncleaned buildings. Data were collected over a two year period from 1997–1999 and included façades cleaned up to 25 years prior to that period. Stone cleaning methods included chemical (acid and alkali/acid combinations), abrasive (grit blasting at various pressures) and high pressure water jetting methods (included under abrasive cleaning). Record keeping with respect to the details of cleaning methods used on façades was poor and acquiring details of

Table 2.1 Number of sandstone building façades surveyed in each city.

City	Edinburgh	Glasgow	Dundee	Inverness
Uncleaned building	28	21	8	8
Cleaned building	36	29	8	8
Total	64	50	16	16

the cleaning methods was one of the most difficult and time consuming aspects of this study. Since detailed cleaning methodologies were generally unavailable, analysis within this study has been confined to splitting the data into three groups – uncleaned, chemically cleaned and abrasively cleaned façades.

The mechanisms by which cleaning can accelerate stone decay are fairly well understood;[19,20,21,22,23] chemical residues can lead to various forms of salt-related decay, abrasive cleaning could open up stone surfaces to increased rates of moisture-related decay. Cleaning may also reduce rates of decay where damaging substances are removed from the stone surface. Stone decay can be related to pollutants deposited in stone[24,25,26] and reduction in their level may reduce the rate of decay. Soiling can form a hydrophobic barrier at the stone surface[27,28] and may block porosity; stone cleaning may therefore reduce rates of decay by reducing moisture levels within stone.

The forms of decay and other alterations to stone building façades quantified during this research are shown in Tables 2.2a and b. In quantifying the surface area of a building affected, stone decay (Table 2.2a) is defined as loss of stone (e.g. granular disintegration) or significant loss of cohesion likely to result in imminent loss of stone (e.g. contour scaling). Areas of previous stone repair or replacement are also counted as 'decay', since these represent previously decayed surfaces. Since this study is concerned with stone decay consequent to cleaning, damage known to have been done by cleaning (e.g. surface roughening by abrasive cleaning) has not been included in the quantification of 'decay'.

Table 2.2a Forms of decay mapped on building façades.

Decay form	Description
Honeycombing	Deep or cavernous pitting in a honeycomb pattern
Case hardening	Hardened crust on top of soft, friable interior
Multiple scaling	Detachment of multiple planar elements parallel to stone surface, unrelated to layering of stone
Crumbling	Detachment of clusters or clumps of grains
Pitting	Pitting of stone surface
Differential decay	Differential weathering rates over stone surface
Decay of components	Loss of clearly bounded elements or inclusions
Granular disintegration	Loss of surface through detachment of individual grains
Flaking	Detachment of small flakes from stone surface
Blistering	Localized blistering of the stone surface
Contour scaling	Detachment of planar elements parallel to the stone surface, unrelated to underlying texture of stone
Dissolution	Partial or selective dissolution of minerals
Back weathering	A single block has weathered back to a significantly greater degree than surrounding stone
Delamination	Detachment of single or multiple planar elements parallel to foliation or bedding plane
Mechanical damage	Loss of compact stone fragments by fracturing
Fissures	Lines of fracture or open cracks through stone
Plastic repair	Mortar repairs to stone surface
Replacement or indenting	Stones that have been partially or wholly replaced or indented

18 STONE DECAY ITS CAUSES AND CONTROLS

Table 2.2b Forms of alteration mapped on building façades.

Alteration form	Description
Soiling	Particulate soiling on stone surface
Black crusts	Thick, black, encrusted soiling deposits
Bird droppings	Bird droppings
Biological growths	Algae, lichens, mosses, higher plants
Colour changes	Bleaching or staining of stone surface
Abrasion from cleaning	Loss detail caused by stone cleaning
Face bedding	Face bedding of layered stone

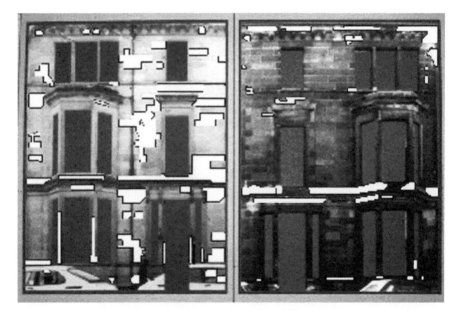

Figure 2.1 Example of rectified images of a pair of cleaned (left) and soiled (right) sandstone building façades. Surface coverage of stone decay and repair is shown as white areas. All areas of façade that are not sandstone (e.g. windows, etc.) have been excluded.

The type and surface coverage of stone decay was determined by observations in the field and mapping of surface coverage of decay onto rectified images of each façade (Figure 2.1, Table 2.3). Details of the methodology used can be found in Ball and Young, Ball et al. and Young et al.[29,30,31] It was not possible to include façade relief features, as only front facing surfaces could be mapped.

Since the façades were of relatively simple design, this is not thought to have significantly reduced the level of accuracy. Buildings were examined from ground level using binoculars where necessary. Data on surface coverage of decay types was digitized in the form of distinct layers on a digital image of each façade using 'Adobe PhotoShop' software (Figure 2.1). Each decay type was digitized separately, but layers can be combined to allow quantification of the overall surface area affected by decay. Using image analysis, the percentage surface coverage relative to the total surface area of stone on each façade can then be calculated.

The mean rate of stone decay on uncleaned façades can be calculated from the age of the building. On cleaned buildings, it is assumed that the façade was left in a sound condition after cleaning and that areas of active decay have occurred since that time. This

Table 2.3 Example of mapped data for façades shown in Figure 2.1.

Façade	1	2
Sandstone type	Bishopbriggs	Bishopbriggs
Cleaned	Chemical	No
Date of cleaning	1985	na
Total decay (%)	17.4	6.7
Total repair (%)	2.0	0.3
Granular disintegration (%)	9.2	4.4
Contour scaling (%)	8.2	2.3
Mechanical damage (%)	0.1	0.0

assumption is necessary since no records were kept of façade condition prior to cleaning. Throughout this paper, rates of stone decay are presented as rates 'per century' although the actual time elapsed since cleaning is normally less than 25 years. Decay rates are averages over a range of time periods, however, decay rates of individual stones and building façades are likely to vary significantly over shorter periods of time.

Data on surface coverage of decay and alteration were transferred to a computer database (4th Dimension) for statistical analysis. The database also included relevant available data on the building façades, including façade age, stone type, date and methodology of cleaning, orientation, etc. In addition to calculating rates of decay and the long-term effects of stone cleaning, this data can be used to estimate future requirements of stone for repair purposes and can be used to provide projections with regard to future repair costs.[32,33]

RESULTS

Comparisons have been made between the rate of stone decay on cleaned and uncleaned building façades. On those façades that have never undergone cleaning it might be expected that older façades would exhibit more stone decay, however, in general, there was found to be no strong relationship between façade age and coverage of stone decay (Figure 2.2). The relationship between stone decay and façade age was complicated by variations in sandstone type within the sample, as this has a significant impact on the rate of decay of individual façades. The mean rate of decay for all uncleaned sandstone façades in the sample was approximately 6% of surface area per century. Decay rates of individual sandstone types may vary substantially from this mean value. The rates of decay for individual sandstone types have also been calculated and are shown in Table 2.4. Rates ranged from 0.5 to 10.5% decay of surface area per century.

Generally, façades are left in a sound condition after cleaning, with decayed stone having been repaired. In this event any current stone decay must have occurred subsequent to cleaning. Figure 2.3 shows, for chemically and abrasively cleaned façades, the relationship between time since cleaning and amount of stone decay (excludes repaired or replaced stone). Again, these data include many varieties of sandstone and consequently there was a large degree of variation. There was an overall trend to increased amounts of decay with time elapsed since cleaning. Calculating regression lines gives values of approximately 85% (abrasive cleaning) and 47% (chemical cleaning) decay of surface area per century. These values represent the rate of stone decay subsequent to cleaning and are approximately an order of magnitude greater than the value obtained for uncleaned façades (approx. 6% decay of surface area per century). This implies that stone cleaning significantly accelerates the rate of decay; however, it is possible that some of the stone decay that occurred after cleaning was a re-activation of pre-existing areas of

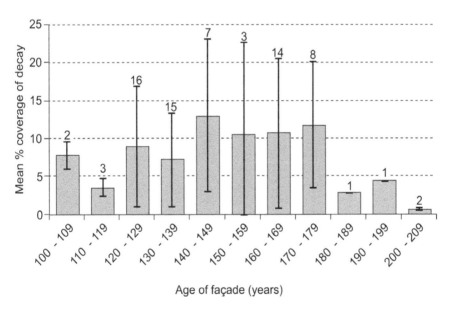

Figure 2.2 Mean % coverage of decay (includes decayed and repaired surface area) on uncleaned building façades aged between 100 and 210 years. Number of façades in each category shown within bar. Standard deviation is shown as error bars.

decay. Although the stone surface may have appeared sound following cleaning, underlying deterioration in the stone structure could result in decay re-appearing within a relatively short time. This would increase the apparent rate of decay following cleaning. Determining to what extent stone cleaning has affected decay over the whole life span of façades therefore requires a more detailed examination of the data.

The preceding data are mean calculations that include a range of sandstones, and there can be substantial differences in the rate of decay between different types. The range of sandstones included within the sample mean that there are relatively few examples of individual stone types. Nevertheless, rates of decay for particular sandstone groups do indicate how varied the response of sandstones to cleaning can be. Table 2.4 illustrates the mean rates of decay per century for particular sandstone types on façades. Where sandstones are petrographically similar, they have been grouped (e.g. Craigleith and Redhall). For both cleaned and uncleaned façades, the rate of decay was calculated over the life span of the building, not the time elapsed since cleaning. As the time since cleaning represents at most 16% of the life span of the façades (the mean is 7%), significant differences between the amount of decay on cleaned and uncleaned

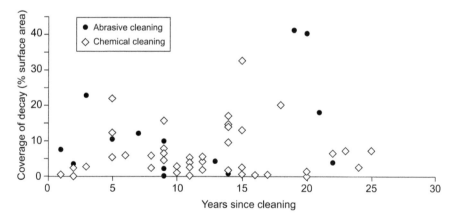

Figure 2.3 Surface coverage of stone decay (excludes repaired or replaced stone) on stone cleaned building façades subsequent to cleaning.

façades may represent a substantial difference in the amount of decay attributable to cleaning. (Results are for paired cleaned and uncleaned façades whose behaviour prior to cleaning can be assumed to have been broadly similar). The rate of decay of individual sandstone types (not cleaned) varied by more than an order of magnitude (0.5% to 10.5% decay per century). In most cases, with the exception of Binny sandstone, more decay was found on cleaned

Table 2.4 Mean rates of decay per century for building sandstones over the life span of the building (100–210 years). NB cleaning took place within the last 25 years. Data show mean % of total surface area of sandstone affected by decay and/or repair. P-values (2-tailed) are for comparisons of pairs of cleaned and uncleaned façades.

Sandstone type	Not cleaned		Chemical cleaning			Abrasive cleaning		
	No. samples	Mean decay (% century^{-1})	No. samples	Mean decay (% century^{-1})	p-value	No. samples	Mean decay (% century^{-1})	p-value
Binny	11	8.1	11	7.6	0.88	2	4.1	0.40
Craigleith & Redhall	11	3.0	11	4.9	0.30	2	2.8	0.94
Locharbriggs	1	0.5	1	0.7	–	1	3.0	–
Moray	10	3.0	3	3.3	0.74	1	9.5	–
Dalmeny	5	3.0	5	7.4	0.01	–	–	–
Leoch & Kingoodie	7	9.2	1	11.5	–	3	22.1	0.01
Bishopbriggs & Giffnock	18	10.5	7	18.5	0.02	7	12.0	0.62

relative to uncleaned façades. Cleaned façades in Binny sandstone had reduced rates of decay relative to uncleaned façades. Some sandstone types, such as Craigleith, Redhall and Locharbriggs, appeared to be little affected by cleaning in terms of their decay rates. Others (e.g. Dalmeny, Leoch and Bishopbriggs) were more strongly affected, with substantially more decay on cleaned façades.

The data in Table 2.4 include decay and repair which have taken place over the entire life span of façades which, while they may be up to 210 years old, were cleaned within the last 25 years. Where stone cleaning accelerates the rate of stone decay it would be most useful to know how much decay took place subsequent to cleaning, however this information is not directly available. Within this research programme, building façades were surveyed from 1997–1999 and records of their condition before cleaning were not available. However, it can be reasonably assumed that prior to stone cleaning the condition of a façade would be similar to that of its paired façade. Where being cleaned or uncleaned is the only difference between the façades, it can be assumed that any difference in the amount of decay and repair between the pairs is largely attributable to cleaning and occurred subsequent to cleaning.

Therefore, for groups of paired façades, subtracting the mean amount of decay and repair on uncleaned façades from the amount on cleaned façades should reveal any excess decay which occurred as a consequence of cleaning (Table 2.5). Vice versa, any decrease in the rate of decay subsequent to cleaning would result in less decay being present on the cleaned façades relative to the uncleaned (negative values on Table 2.5).

Results (Table 2.5) indicate that, for many façades, stone decay was accelerated after cleaning. Mean values were 42% (abrasive

Table 2.5 (opposite) Mean rates of decay per century for building sandstones. Uncleaned façades show mean rate over building life span. Cleaned façades show the calculated rate after cleaning. Data shown are mean % of total surface area of sandstone affected by decay and/or repair. Negative values for decay after cleaning indicate that the amount of decay on cleaned façades was less than that on uncleaned façades.

QUANTIFICATION OF THE DECAY RATES

Sandstone type	No. façade pairs	Uncleaned Mean decay (% century^{-1})	Standard deviation	Chemical cleaning Mean decay (% century^{-1} after cleaning)	Standard deviation
Craigleith & Redhall	11	3.9	4.3	54	207
Binny	12	6.2	5.2	25	74
Dalmeny	6	2.8	0.8	63	31
Moray	3	2.6	1.6	0	23
Locharbriggs	1	0.5	nd	1	nd
Leoch & Kingoodie	1	6.5	nd	48	nd
Bishopbriggs & Giffnock	8	11.8	7.4	156	505
Silverstone	1	0.2	nd	39	nd
Mean				61	238

Sandstone type	No. façade pairs	Uncleaned Mean decay (% century^{-1})	Standard deviation	Abrasive cleaning Mean decay (% century^{-1} after cleaning)	Standard deviation
Craigleith & Redhall	2	1.3	1.4	37	65
Binny	2	17.1	3.9	-ve	27
Dalmeny					
Moray					
Locharbriggs	1	0.5	nd	34	nd
Leoch & Kingoodie	3	8.3	3.3	105	84
Bishopbriggs & Giffnock	7	9.3	5.1	57	164
Silverstone					
Mean				42	129

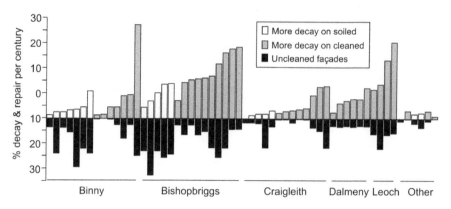

Figure 2.4 Rate of decay and repair per century (% surface area) for individual uncleaned and cleaned (chemical and abrasive) pairs of façades.

cleaning) and 61% (chemical cleaning) decay of surface area per century subsequent to stone cleaning. For most sandstone types the mean rate of decay accelerated following cleaning, however, there was substantial variation within each group. Figure 2.4 shows the degree of variability in response within each sandstone type by displaying the data for individual façade pairs. In most cases there was more decay on the cleaned façade of each pair; however, some cleaned façades suffered less decay than their uncleaned pair. This indicates how variable the response to stone cleaning can be, even within the same sandstone type. It is likely that this is due to differences in cleaning technique, in the details of the method used and degree of care taken in cleaning. Details of individual cleaning methods are normally unavailable as records are seldom kept. It is therefore not possible to relate variations in rate of decay to variations in cleaning method.

On average, the rate of stone decay subsequent to cleaning has been accelerated by approximately an order of magnitude in the years immediately following cleaning. It appears that decay rates are initially accelerated immediately following cleaning and later decline to more 'normal' levels. This is illustrated in Figure 2.5; the y-axis shows, for individual cleaned façades, the calculated rate of decay per century subsequent to stone cleaning. It is clear that rates of

QUANTIFICATION OF THE DECAY RATES 27

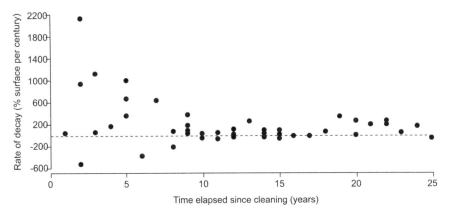

Figure 2.5 Relationship between time elapsed since cleaning and the calculated rate of decay after cleaning (abrasive and chemical) for individual façades. Data points below the axis are façades where the decay rate after cleaning was less than that before cleaning.

decay decline with increasing time elapsed since cleaning. Since any decay caused by cleaning will remain on the façade, then on average the calculated rate of decay should, and does, always remain higher than that for uncleaned façades. Figure 2.5 also contains data (plotting below the x-axis) from façade pairs where the cleaned façade had less decay than the uncleaned façade. The distribution of this data implies that, where reduced rates of stone decay occurred following cleaning, the rate then increased again as time elapsed after cleaning.

The preceding results imply that the effects of cleaning on rates of stone decay are most pronounced in the decade after cleaning. Given the extremely accelerated rates of stone decay that have been shown to occur during this period, it is clear that cleaning has the potential to significantly increase the long-term amount of decay on building façades. In terms of their rate of decay, some sandstone types generally responded badly to cleaning, others may be affected only slightly and some had decreased rates of decay for a period. Although it was generally true that stone cleaning resulted in increased stone decay, it is clear that variations in cleaning method and stone type substantially affected the outcome of cleaning.

DISCUSSION

Since this study was aimed at predicting façade maintenance requirements and estimating the future need for stone for repair purposes[34] it has concentrated on estimating the rate of stone decay in relation to progression in its surface coverage on façades. Stone decay can also progress by increasing depth of penetration into stone and by increasing friability within one location. This study does not attempt to address these forms of stone decay. Since there is no available information about variations in rate of decay over time, the rates calculated here represent average decay rates. In reality, decay rates are unlikely to be linear[35,36,37] and the results in Figure 2.5 imply that, for cleaned façades at least, the rate of decay can change significantly within a few years. It is not possible to examine short-term variations in rate of decay with the data gathered in this study.

Examination of results in terms of the coverage of stone decay and the age of uncleaned façades found no strong relationship between these variables. This was partly caused by the range of sandstone types included in the sample, but it is also possible that façades exhibiting more decay were more likely to have been demolished, removing them from the sample. In this case, the estimate of mean rate of decay for uncleaned building façades (6% surface area per century) would underestimate its true rate. Clearly the rates of decay of different sandstone types can vary substantially. Rates for sandstones included in this study varied from approximately 0.5 to 10.5% decay of surface area per century. These underlying rates of decay can also be an indicator of the likely response to cleaning as sandstones with a higher 'natural' rate of decay tended to be those where decay was most accelerated after cleaning. However, this was not always the case; Binny sandstone had a relatively high 'natural' rate of decay, but its rate of decay after cleaning could be reduced for a time. Façades displaying reduced rates of decay following cleaning were in the minority. They included both abrasive and chemical cleaning methods and a variety of sandstone types. There was insufficient detail available on cleaning methods and stone condition to determine the reasons for this response within this study; potential mechanisms include reduction in pollutant levels and increased permeability resulting in reduced moisture loading.

Although rates of stone decay were sometimes reduced for a period, the more common effect of cleaning was to cause increased rates of stone decay. While the mean 'natural' rate of decay on uncleaned façades was approximately 6% surface decay per century, mean rates on cleaned façades were an order of magnitude greater. The mean rate for abrasively cleaned façades was calculated to be approximately 42% surface decay per century. The mean rate for chemically cleaned façades was approximately 61% surface decay per century. However, the response of individual façades can vary greatly from these means and the effects of stone cleaning on rates of decay appear to decline over time.

The rate of stone decay in the years immediately after cleaning is not linear. Both initially increased and initially decreased rates of decay were at a maximum immediately following cleaning, declining over the following decade or so. This decline does not imply that the results of stone cleaning become negligible over time. Where decay is accelerated by cleaning, the excess decay during this period will remain on the façade and the overall amount of decay on such a façade will always remain higher than that on the equivalent uncleaned façade. On sandstones prone to rapid decay, cleaning has the potential to significantly increase the total amount of stone decay.

Although in some circumstances rates of stone decay may be reduced following cleaning, this should not be taken as a recommendation to clean particular stone types. It has been shown that the effects of cleaning methods can be highly variable and it is possible to cause accelerated decay by the use of inappropriate cleaning methods. In addition, the preceding results take no account of other forms of damage that may be caused by cleaning, including surface roughening, abrasive loss of detail, bleaching and staining. Binny sandstone, for instance, is very vulnerable to colour changes following cleaning. Such effects may be considered as aesthetically damaging as increased levels of stone decay.

These results illustrate the ongoing effects of stone cleaning carried out over a 25 year period. Cleaning methods vary in their effects on stone and, as a consequence of previous research[38,39,40] and product developments, the methods themselves have undergone significant changes over this period. The most aggressive methods of the past (e.g. high pressure grit blasting and highly concentrated chemical cleaning agents) are now seldom employed. The post-cleaning behaviour of recently cleaned façades and those cleaned in the future may therefore differ from that of similar façades cleaned by previous, more aggressive methods.

Acknowledgements

This research was commissioned and funded by Historic Scotland.

References

1. Webster, R.G.M., Andrew, C.A., Baxter, S., MacDonald, J., Rocha, M., Thomson, B.W., Tonge, K.H., Urquhart, D.C.M. and Young, M.E. (1992) *Stone cleaning in Scotland. Research Report.* Vols. 1–3, Report to Historic Scotland and Scottish Enterprise by Masonry Conservation Research Group, Gilcomston Litho, Aberdeen.
2. Andrew, C.A., Young, M.E. and Tonge, K.H. (1994) *Stone Cleaning: A Guide for Practitioners.* Historic Scotland and The Robert Gordon University.
3. MacDonald, J., Thomson, B. and Tonge, K.H. (1992) Chemical cleaning of sandstone – comparative laboratory studies. In Webster, R.G.M. (ed.) Stone cleaning *and the Nature, Soiling and Decay Mechanisms of Stone.* Proceedings of the International Conference, 14–16 April 1992, Edinburgh, UK, Donhead Publishing, London: 217–226.
4. Nord, A.G. and Tronner, K. (1992) Characterization of thin black layers. In Delgado Rodrigues, J., Henriques, F., Telmo Jeremias, F. (eds.) *7th International Congress on Deterioration and Conservation of Stone.* 15–18 June 1992, Lisbon, Portugal: 217–225.

QUANTIFICATION OF THE DECAY RATES 31

5   Jones, L.D. (1986) Criteria for a selection of a most appropriate cleaning method. In Clifton, J.R. (ed.) *Cleaning Stone and Masonry*. American Society for Testing and Materials, Philadelphia: 52–67.
6   Webster *et al, op. cit.* (1992)
7   Andrew *et al, op. cit.* (1994)
8   Werner, M. (1989) Research on cleaning methods applied to historical stone monuments. In Baer, N.S., Sabbioni, C. and Sors, A.I. (eds.) *Science, Technology and European Cultural Heritage*. 13–16 June 1989, Bologna, Italy: 688–691.
9   Werner, M. (1991) Changes to the surface characteristics of sandstone caused by cleaning methods applied to historical stone monuments. In Baker, J.M., Nixon, P.J., Majumdar, A.J. and Davies, H. (eds.) *Proceedings of the 5th International Conference on the Durability of Building Materials*. 7–9 November 1991, Brighton, UK: 225–234.
10 Bluck, B.J. and Porter, J. (1991) Sandstone buildings and cleaning problems. *Stone Industries* March 1991: 21–27.
11 MacDonald *et al, op. cit.* (1992)
12 Webster *et al, op. cit.* (1992)
13 Urquhart, D.C.M., Young, M.E. and Cameron, S. (1997) *Stone Cleaning of Granite Buildings*. Historic Scotland Technical Advice Note 9, The Stationery Office, Edinburgh.
14 Urquhart, D.C.M., MacDonald, J., Thomson, B.W., Tonge, K.H. and Young, M.E. (1993) Stone cleaning in Scotland: A Cause for Concern? In *Management, Maintenance and Modernisation of Buildings*. Proceedings of the International Symposium, Vol. 6, CIB Working Commission 70, Rotterdam, 28–30 October, 1992.
15 Urquhart, D.C.M., MacDonald, J., Tonge, K.H., Webster, R.G.M. and Young, M.E. (1994) Sandstone cleaning: a risk assessment. In *Building Pathology 92*. International Conference on Building Pathology, 23–25 September 1992, Cambridge.
16 Young, M.E. and Urquhart, D.C.M. (1998) Algal growth on building sandstones: effects of chemical stone cleaning methods. *Quarterly Journal of Engineering Geology* 31: 315–324.
17 Young, M.E. (1998) Algal and lichen growth following chemical stone cleaning. *Journal of Architectural Conservation* 4(3): 48–58.
18 Young, M.E., Ball J., Laing, R., Scott, J. and Cordiner, P. (2001) An investigation of the consequences of past stone cleaning intervention on future policy and resources. Report to Historic Scotland.
19 Werner, *op. cit.* (1989)
20 Werner, *op. cit.* (1991)
21 Bluck and Porter, *op. cit.* (1991)
22 MacDonald *et al, op. cit.* (1992)
23 Andrew *et al, op. cit.* (1994)

24 Smith, B.J., Whalley, W.B. and Magee, R. (1989) Building stone decay in a 'clean' environment: western Northern Ireland. In *Science, Technology and European Cultural Heritage, Proceedings of the European Symposium.* 13–16 June 1989, Bologna, Italy: 434–438.
25 Sabbioni, C. and Zappia, G. (1992) Decay of sandstone in urban areas correlated with atmospheric aerosol. *Water, Air and Soil Pollution* 63: 305–316.
26 Halsey, D.P., Mitchell, D.J., Dews, S.J. and Harris, F.C. (1995) The effects of atmospheric pollutants upon sandstone: evidence from real time measurements and analysis of decay features on historic buildings. In *10th World Clean Air Congress.* 28 May–2 June 1995, Espoo, Finland.
27 Bluck and Porter, *op. cit.* (1991)
28 Nord and Tronner, *op. cit.* (1992)
29 Ball, J. and Young, M.E. (2000) Mapping the decay and weathering of stone: A technique for the assessment of large numbers of buildings. In *Proceedings of the New Millennium International Forum on Conservation of Cultural Property.* 5-8th December 2000, Deajeon, South Korea, pp. 134–147.
30 Ball, J., Young, M.E. and Laing, R.A. (2000) Rapid field assessment of stone decay on buildings and monuments. In Fassina, V. (ed.) *9th International Conference on Deterioration and Conservation of Stone.* Vol. 2, Venice, 19–24 June 2000, Elsevier, Amsterdam: 13–21.
31 Young *et al, op. cit.* (2001)
32 Laing, R.A., Al-Hajj, A., Ball, J., Scott, J. and Young, M.E. (1999) Stone cleaning: A life cycle cost model. In Lacasse, M.A. and Vanier, D.J. (eds.) *Durability of Building Materials and Components 8.* Vancouver, Canada, 30 May–3 June 1999, NRC Research Press, Ottawa: 1739–1745.
33 Laing, R.A., Ball, J., Scott, J. and Young, M.E. (2000) The implications of stone cleaning for planned building maintenance. In Fassina, V. (ed.) *9th International Conference on Deterioration and Conservation of Stone.* Vol. 2, Venice, 19–24 June 2000, Elsevier, Amsterdam: 813–817.
34 Young *et al, op. cit.* (2001)
35 O'Brien, P.F., Bell, E., Orr, T.L. and Cooper, T.P. (1995) *Stone loss rates at sites around Europe. The Science of the Total Environment.* 167: 111–121.
36 Kohler, W. (1996) Investigations on the increase in the rate of weathering of Carrara marble in Central Europe. In Riederer J. (ed.) *Proceedings 8th International Congress on Deterioration and Conservation of Stone.* 30 Sep–4 Oct 1996, Berlin, Germany: 167–173.
37 Smith, B.J. (1996) Scale problems in the interpretation of urban stone decay. In Smith, B.J. and Warke, P.A. (eds.) *Processes of Urban Stone Decay.* Donhead Publishing, London: 3–18.
38 Webster *et al, op. cit.* (1992)
39 Andrew *et al, op. cit.* (1994)
40 Urquhart *et al, op. cit.* (1997)

3 Durability and Rock Properties

R.J. INKPEN, D. PETLEY and
W. MURPHY

ABSTRACT

Durability is a complex term that can only be defined and understood in relation to its context. Different contemporary attempts to define durability have concentrated upon the physical aspects of the term, such as stone properties, although many have noted and acknowledged a human aspect of varying degrees of significance. Analysis of the context of the development of the salt crystallization test at the BRS in the 1920s and 1930s illustrates how durability was constructed from the network of individuals, equipment and stone within a specific social and economic context. Assessment of durability and the role of stone properties within these tests cannot be understood apart from this complicated set of relationships.

INTRODUCTION

Rock durability is a major concern within construction and conservation work. Determination and classification of rocks according to durability can result in their specification or lack of it for building projects and restoration work. Within the United Kingdom, the quasi-standard of the salt crystallization test for assessing durability has a long history of use and criticism.[1] Sedman and Stanley, for example, criticize the limited sample size used for the determination of durability[2,3] whereas Ross and Massey provide a contrary opinion.[4] Within the European Union a range of standards for assessing the durability of different rock types are being established by discussion and comparison of national practices and standards.[5,6] All the tests being considered or already agreed rely on the use of laboratory testing of samples for the determination of durability, often using a single index of alteration as the criterion for assessment. Likewise, recent academic analyses of durability, such as Goudie and Moh'd *et al.*,[7,8] concentrated upon durability as a phenomenon discernible by laboratory assessment of identifiable and quantifiable natural material properties. This short paper suggests that the view of durability as predictable from laboratory analysis of the relationship between a small number of properties may provide an extremely limited view of a complex and multi-faceted phenomenon. Initially this paper focuses on the definition of durability, and the integral role of human use and workmanship in defining and determining durability is highlighted. Laboratory studies also demonstrate that the contextual nature of the properties measured and tests run make the extension of such results as a general characterization of durability difficult to maintain. The significance of context for defining, or rather constructing, durability is illustrated through an analysis of the development of the salt crystallization test at the BRE in the 1920s and 1930s. The conceptual basis for the testing is outlined and it's relationship to context discussed. Finally, the pivotal role of Portland Stone in defining an acceptable test is then illustrated.

DEFINING DURABILITY

Turkington points out that durability is a vague and ambiguous term often being allocated definitions that depend on the field of the researcher.[9] Turkington suggested that the term embraces a number of individual weathering processes and stone properties making a precise scientific study of it difficult.[10] By concentrating on the physical aspects of the term, she identifies durability as being synonymous with the nature and rate of weathering processes on exposed rock, the alteration of material and the creation of secondary weathering products. The Building Research Establishment viewed durability as being based upon the physical characteristic of the stone and the nature of the cementing material.[11] In the United States a similar view of durability, as related to innate characteristics of the stone, was also put forward,[12] but tempered by recognition that environment and use of the stone would influence the determination of durability. Carr *et al.* viewed differences in mineralogy and chemical and physical properties of stone to be too great to have a single test to evaluate durability.[13] They suggested that proven use over time is the best test of durability. In the United Kingdom, by 1997, whilst the rate and form of weathering was still viewed as a function of the internal characteristics of the stone, the BRE also explicitly recognized that durability was complex. Durability also depended upon the location in which the material was placed and upon the relationships between the material and the rest of the building. The complex relationships this definition entailed meant that a single durability test was no longer seen as of universal application.[14] Similarly, in the United States durability is viewed as a complex phenomenon.[15,16]

Viewing durability as a property inherent within a stone implies that stone selection is a task reducible to measurement of specific properties at a single point in time. Given a set of properties, durability follows as a direct measurement and is often viewed as synonymous with strength.[17] This view negates the need to view durability as a term defined by relations both physical and social rather than a term with a fixed definition. Garden, for example, viewed durability as the outcome of the interaction of material and environment.[18] Included within this definition were the in-service expectations for the material such as the locations of exposure. Nireki similarly, saw performance evaluation over time as a vital part of durability.[19] As this method required comparison of performance against performance requirements, it rejected the view of durability as discernible by a test for a single property of an object. Performance implied that the stone is under constant assessment in relation to the expectation of some designed function. Durability is not a static initial value, but a continually adjusting assessment of the state of the stone. Duffy and O'Brien again highlight the significance of performance for defining durability.[20] They pointed out, however, that performance is different for different interest groups such as architects and engineers. They settle on a definition of durability as the resistance of a stone to changes in its physical, mechanical and aesthetic properties over time. This definition incorporates both measurable scientific elements and a social/cultural factor in the aesthetic element. For both there is a potential change over time. Within their study a key role is played by the surface environment as determinant of the supply of weathering agents and range of weathering processes. As the surface environment changes so the physical properties alter and hence durability changes.

The human aspect of durability has often been ignored or deemed irrelevant to scientific study. Schaffer noted that workmanship could have as great a significance upon stone degradation as the physical properties of the stone.[21] Likewise, pamphlets on stone selection[23,23] highlight the importance of the preparation and working of the stone for its final weathering behaviour. Similarly, conservation treatments such as limewashing can have differential impacts upon the same stone type depending upon the expertise with which they have been applied.[24] Both these aspects highlight the interplay between the usually starkly defined physical and human parts of the weathering system. Any stone used in a building, its precise location, its form and its relationship to other parts of the building are all parts of a human design, whether by a known architect or unknown builder. Each decision in the initial building and subsequent maintenance is part of a human system that interacts and is informed by the nature of the building materials available. Isolating and studying one part of the whole system provides only a partial picture of durability and how it is defined and operated upon. At Exeter Cathedral in the early nineteenth century, for example, replacement of susceptible stonework, such as pinnacles, was based as much on its visible deterioration and aesthetic impact as upon the reports of an expert surveyor.[25] Durability was defined as much by management issues within the Cathedral as by physical properties of the stone.

At another scale, durability and its definition cannot be divorced from the wider socio-economic system within which it is being defined. Within contemporary society, for example, defining durability could be viewed as being intimately tied to the general commodification of natural products within the economy.[26] Lewry and Crewdson in a study of approaches to testing durability state five key players interested in stone durability; the client, architect, contractor, supplier and manufacturer.[27] The relationship between the client and architect (or specifier) is seen as the central driving force in developing durability testing. The architect in particular is acutely aware of the problems of potential litigation if stone selection results in aesthetic or structural problems. Establishing a standard test which stone has to pass provides a point of reference for the architect and shifts responsibility for the problem to the test rather than the selector. The other players in this group have economic and social interests in the development of such testing. Likewise, the definition by measurement of a limited number of properties in a standard manner could be viewed as a means of ensuring that durability, like other standards, becomes amenable to management, intervention and exploitation.[28]

Turkington suggested that geomorphological views of durability converge,[29] despite their distinct disciplinary objectives, with other groups interested in stone degradation. The focus of convergence is the prediction of the nature and rate of deterioration of stone. Whilst there is some concern with the contextual aspects of deterioration, such as the location of stone and its microenvironment, most studies in this mould have been inherently reductionist. Geomorphologists, and to a great extent organizations in search of standards, have tended to concentrate upon characterizing stone and simulating weathering environments within the laboratory.[30] The complex weathering system that is part of durability is reduced to a set of simple, measurable relationships. The assumption seems to be that isolating and identifying specific interactions between measured properties and measured inputs under controlled conditions results in a deep understanding of how the stone degradation system operates at the scale of a whole building. The small-scale laboratory studies are merely miniature versions of the larger buildings and relationships can be extrapolated up to the building scale.

CONSTRUCTING DURABILITY: EARLY TESTING AT THE BRE

Experimental work is often seen as deriving truths about the real world via controlled, ordered and repeatable investigation. Experimental work is an intervention into the real world. Phenomena are theorized and caged within a controlled framework to try to extract their effects in a regular manner.[31] The effects are not viewed as constructed by the observer, but are seen as being out there in reality awaiting illumination by the experimenter. Such effects may not, however, exist anywhere outside of the experiment within which they are created. In reality complex interactions entangle experimentally isolated phenomena. As observers we cling to the idea that the effect is still there, maintaining an explanatory role for it, but enhanced or inhibited by other factors. In so doing we are imposing a theoretical construction from the laboratory onto reality in which the potential explanatory role of the complexity of interactions is ignored.

Experimental assessment of durability could be viewed as such a construction. Understanding why and how such tests were designed could be of use in exploring and explaining the relationship between experimental work and durability on buildings. Analysis of the development of the salt crystallization test at the BRE (then the BRS, the Building Research Station) during the 1920s and 1930s illustrates the significance of inter-related factors. Central to test development were the motivations of researchers, the context within which the work was carried out including laboratory facilities, as well as the important role of the stone itself in the process of interrogation. All references to this work are derived from the archives of the BRE and specifically files 42/5, 43/1, 43/2, 44/1 A3.W1.E700. Viewing the experimental work as part of a complex and evolving network of actors can help to direct analysis in understanding experiment construction. Likewise, any individual experiment can be viewed as a network in its own right directing and ordering other actors, including materials, to produce an outcome that becomes part of the evolving larger network of the laboratory itself. Viewed in this way no experiment can be understood in isolation from the relations that form the network of which it is a part. Work on the role of social influences at a range of scales upon laboratory work and science in general have been developed by authors such as Latour and Woolgar[32], Latour,[33] Callon[34] and Law.[35] They have suggested that such complex networks of relationships can be found in a variety of scientific settings and are of relevance in interpreting results from laboratories. Deremitt within climatic research has interpreted his results as at least partially socially constructed.[36] Whilst it is not the objective of this paper to outline this approach in detail some elements of it will be used in the analysis of the development of durability testing at the BRS and the role of rock properties in this development.

Durability testing began at the BRS in June 1927, a short time after the first meeting of the Select Committee on Stone in 1926. The approach of the BRS to the problem of durability is outlined in the draft of a station bulletin that they were instructed to prepare in September 1928. Although very tentative in tone, this report highlights the conceptual basis upon which future testing would be carried out. This included the assumptions about what durability was and how it could be scientifically and consistently measured. The first section of the bulletin sets out the problem as seen by the BRS. The report made a distinction between scientific experience, as a trained observer and investigator, and practical experience of stone masons and architects. Practical experience as a guide to durability is seen as useful, but not applicable to stone from a new quarry or a reopened old quarry. Significantly, practical experience is viewed as of little value in comparing building materials whether artificial or natural. Such information is seen as obtainable economically, quickly and with less risk by the application of scientific methods. To what risk the report was referring is not made clear. It can be assumed, however, that economic concerns were a major determinant of risk in this context. The devaluing of practical experience as a guide to durability is linked to the needs of the economy and society for building material that can be characterized rapidly and used with a feeling of certainty in its performance. The development of durability tests cannot be divorced from the social and economic context:

> It is not suggested that laboratory investigation can dispense with practical experience. Experience is invaluable, but under modern conditions, practical experience is unable to supply all the information required, in the time available.
> File A3.W1.E700.

In an echo of this concern with modern needs of efficiency Moh'd et al. suggested that their experimentally derived relationship between saturation and porosity and assignment to BRE durability classes can be done in hours rather than in 3–4 weeks.[37]

Durability is still viewed as difficult to define and study via scientific methods, a point Schaffer emphasized again in 1932. The issue was seen as so complex that no attempt is made to estimate the life of the stone in terms of years. Instead the life of a stone was seen as being an indeterminate quality that depends upon aesthetics and structural factors. It was considered important, however, not only to consider rates of weathering, but also the type of decay to which a material was liable. From the start of testing the type of decay was seen to be as important as the rate and often the determinant of whether decay had occurred. Despite the recognition of the importance of environment for durability, the power of resistance of stone was viewed as dependent upon its physical and chemical properties alone. The view of the stone as resisting decay set it up as a static entity in opposition to the dynamic environment in which it decayed. On this basis properties could be viewed as fixed from the time of exposure. Durability therefore becomes a characteristic that could be determined and remain constant, or at least consistent, for any stone, fresh or weathered.

Testing of stone was divided into three types: accelerated weathering, the empirical method and the comparative method. Accelerated weathering was viewed as the worst option of the three as there is no way to ensure that making conditions more drastic than those in reality reproduces the effects of decay at an increased rate. The empirical method measured stone properties that had been observed and known from experience to have an effect upon durability. Although this approach was viewed as scientific, it suffered from the inability to measure the properties of interest, which are never made explicit, nor their direct impact upon weathering behaviour. The comparative method is the one that the BRS decided to be most appropriate. In this method the behaviour of stone with known properties and weathering behaviour is compared to the properties and behaviour of an unknown material. The basis for comparison relies upon having, firstly, sufficient information on the weathering behaviour of materials and physical and chemical data on them and, secondly, the supervision of an investigator of sufficient experience to balance justly the results of the tests. The latter are required as different test conditions will be of differing significance for different stones. Similarly, the experienced investigator is essential for interpreting the results of any tests. Test results only reflect the susceptibility of a stone to a single agent under standard conditions. The significance of weathering forms produced and whether the processes are producing 'real' weathering are all within the power of the experienced investigator to decide. The investigator, therefore, becomes a central figure in determining the significance of the test results and their relationship to reality. Rather than a single, transferable number to represent durability, durability becomes an estimation made by the investigator.

The importance of economic context for the development of the durability test is also visible in the procedure for the selection of stone for testing. Representativeness is seen as a problem, but the ideal solution is viewed as the purchase of stone in the open market. It was stone as a commodity and the representative nature of samples relative to that definition that was the key concern for the BRS, not stone as a natural material. Once in the market place, stone had to be treated as any other commodity. Buyers may have no knowledge of the product and need not be local to the source of the stone. Information that had been adequate when stone was used locally was not adequate for stone as a trade good. Testing could, however, be informed by local knowledge. Examination of buildings near the quarry could help to characterize the type of decay experienced, but less weight was given to these buildings as indicators of durability. Specific mention was made of the rural location of most quarries and the inappropriateness of these locations for assessing durability in urban environments. Urban areas would be where stone was purchased and used; without detailed local knowledge of how the stone performed, this is where there is most need of standard characterization.

A range of tests for comparative testing were investigated, but the salt crystallization test was considered to be the most valuable. Crystallization was viewed not as a simulation of frost weathering as originally by Brard,[38] but as the predominant cause of decay. Propensity towards this decay mechanism was also viewed as providing a correlation with the size and distribution of pore spaces and therefore with structural factors that were considered impossible to measure at the time. This was seen as a major strength of the test. Likewise, initial test runs, probably in June 1927, suggested that the test produced the type of decay that had been observed on buildings. It is at this point that it is explicitly stated that the type of weathering is more significant that the rate, thereby enhancing the value of this test as an accurate simulate of weathering. In particular, changes to the sample throughout the test can be monitored by the investigator to check that they match reality. A developmental sequence of expected changes is assumed by the investigator and the test must match this to be accepted as a valid representation of reality.

Although no documents concerning the June 1927 tests could be found in the archives they are continually referred to in the early tests in 1929–1932. The 1927 tests are taken as the benchmark set of tests against which the systematic test series 1–52 should be compared (File 43/2). The criteria against which unsatisfactory tests and the proposed test series were assessed was not reproduction of weight change, the variable measured, but the decay forms produced. Despite the continued reference back to Brard,[39] it was a paper by Orton in 1911 that formed the basis for the experimental design employed.[40] Orton discussed ceramics rather than stone in his test procedure. The readiness with which tests developed for other materials were translated for use on stone is indicated by the continual reference to stone in most Bulletins (such as A3.W1.E700). This reflects a view of stone as a homogeneous commodity capable of standard testing. This assumption means that it can be treated as a material assessable by methods used on materials such as ceramics with only minor alterations to procedure.

Concentrating on reproducing weathering forms on a few stone types, particularly Portland Stone, limited the type of weathering forms that could develop as well as the types of alteration that could be made to the experimental design. The visible forms of blistering and cracking found on Portland Stone buildings by the BRS in the 1920s became the only forms of weathering of concern within the tests. Dissolution, for example, was not considered as a form of weathering within the test series. Testing for durability relative to this form of decay was not seen as important. The visually striking decay caused by salts became the focus and definition of durability.

Test series 17–22, 35 and 42 highlight the importance of Portland Stone for defining what is acceptable as weathering within the test. In total 166 samples were collected during a survey of Portland quarries and tested in different series. The weathering form was consistent between samples although some decayed more rapidly than others. Decay began first at the edges and spread rapidly over the whole surface producing uniform decay. Additionally, two other forms of weathering were accepted as typical of Portland Stone. In the first, the edges and corners decayed rapidly and a spherical form was produced, whilst in the second, decay was concentrated in patches. These forms of weathering on small 40 mm cubes became the standard type of weathering expected and deviations from this required explanation. Likewise, the survival of samples through the 15 crystallization cycles was assumed to be standard as well. The interaction of Portland Stone with the experimental conditions had therefore constructed a rigid set of practices that were thought to replicate stone decay forms on buildings.

The significance of Portland Stone for defining the standard test procedure can also be seen in test series 35 and 37. In series 35 Portland Stone was compared to the decay of French limestones and magnesium limestones along with various other stone types. Although the results were deemed satisfactory, deviations in the form of weathering for a few samples were explained as aberrations. A sample of magnesium limestone had a clayey layer which concentrated decay there, whilst another was explained as decaying rapidly because of its lower saturation coefficient. The disintegration of all the French limestones by the tenth cycle was not viewed as a problem either. Instead the physical characteristics of these limestones were described and their poor performance attributed to these. The implication is that the properties differed from those of the standard Portland Stone and so variability in performance was explained. By implication this performance had to be repeated in reality so these limestones were deemed to have an absolute lesser durability.

By series 35, the test itself is seen as an infallible measure of durability, any differences in weathering forms from those expected from the standard are seen as a problem, not a focus for investigation. In series 37 Bath Stone and French limestones were subjected to the standard test, but using only 5% sodium sulphate solution. The write-up for the whole test covers just one side of paper. The samples survived the whole test run and decayed in a slightly different manner from the Portland Stone samples. The results were not followed up, which suggests that devising a range of crystallization tests sensitive to each type of stone was not considered. This illustrates how the investigators had come to view the weathering forms produced by the standard test as the only forms of interest and the test itself as an accurate and real representation of durability.

By series 48, a rigorous comparison between physical properties and decay performance was being sought. Portland Stone was vital for constructing any relationship between physical properties and weight loss. The other stone types used in this series did not plot along the same general trend identified by the Portland Stone samples (Figures 3.1 and 3.2). The crystallization test was thus designed to reflect a belief about the dominant decay forms on Portland Stone and that belief is reflected in the association displayed in the graph. Correlation coefficients were calculated for Portland Stone data from this test. The strongest correlation (-0.87) was found between weight loss and saturation coefficient of the stone implying that the availability of pore spaces was a key parameter in the effectiveness of the salt treatment applied. More detailed data on microporosity was also obtained at the time for a subset of the samples in series 48. Regressing the data illustrates the importance of Portland Stone for establishing that there was a relationship between rock properties and weight loss (r^2 values of 0.844 for saturation coefficient, 0.342 for % porosity and 0.842 for % micro-porosity). In the case of the Portland Stone

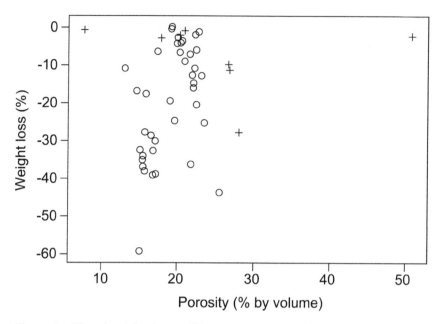

Figure 3.1 Plot of weight change (%) against porosity (% by volume) for all samples in crystallization series 48. Portland Stone samples (O), other stone types (+).

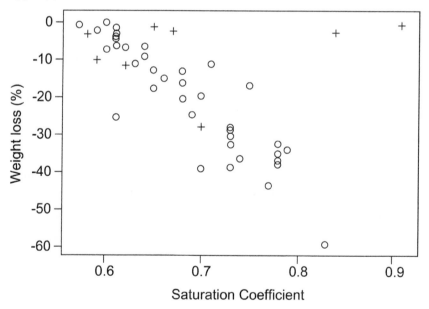

Figure 3.2 Plot of weight change (%) against saturation coefficient in crystallization series 48. Portland Stone samples (O), other stone types (+).

data it is also worth noting that the samples seem to split into two clusters. The relationship between porosity and weight loss is not as strong as that between weight loss and the saturation coefficient. This could be the result of the closeness of a number of samples in terms of their porosity between 15% and 16% that exhibit similar weathering behaviour. The dispersal of saturation coefficient values is greater and such limited behaviour for samples with similar saturation coefficients is not observed. This suggests that the saturation coefficient provided, as measured at the time, was a much better indicator of differences in the experimental weathering behaviour to be expected from Portland Stone samples.

All the Portland Stone samples in series 48 were obtained from existing buildings. Each was graded according to its conditions and, as Figure 3.3 illustrates, there is a good agreement between weight loss, saturation coefficient and the condition of the stone assigned by the experienced investigator. The good association is not unexpected

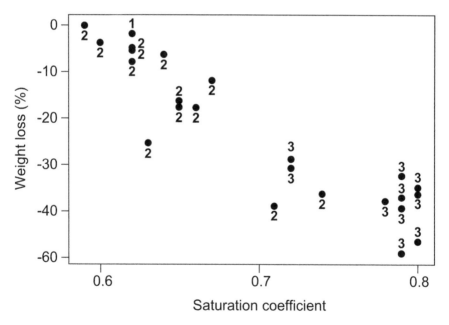

Figure 3.3 Plot of Portland Stone samples from series 48 labelled by quality of stone as assessed by BRS in 1920s. (1) excellent condition; (2) good condition; (3) bad condition.

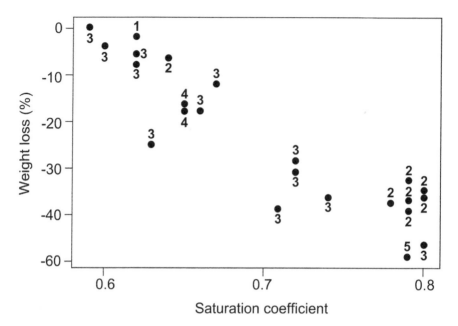

Figure 3.4 Plot of Portland Stone samples from series 48 labelled by the building from which the stone was obtained. (1) St. Paul's Cathedral; (2) Victoria Memorial; (3) Williams Deacon Bank; (4) Mansion House; (5) Ministry of Works.

given that the test was structured to produce a particular response from Portland Stone samples. Combining this information with data on buildings the samples come from (Figure 3.4) suggests that some of the relationship may result from the stones being from the same building as well as being in the same condition. Former weathering of the samples also produced the appearance of decay and altered rock properties. Alteration does not, however, seem to be consistent between buildings. The trends are not repeated once samples from the same buildings are analysed except in very general terms. Significantly, the investigators did not feel that including data from weathered blocks was a problem in determining durability. The view of durability as a single and constant or consistent measure meant that weathered stone could be expected to retain and exhibit initial durability properties despite being weathered. Durability was seen as a property inherent in and retained by the stone, not a property that developed through interaction with the environment.

CONCLUSIONS

Analysis of the development of the salt crystallization test suggests that the test cannot be understood in isolation from the individuals who constructed it and their initial view of the importance of the decay forms they were trying to produce. The role of the investigator in deciding on the outcome of a test is central to developing an assessment of durability. Similarly, Portland Stone becomes the standard rock against which all other rocks have to be compared. This is not to say that the test is a fabrication with no contact with reality. The response of Portland Stone, the equipment and all other physical entities are real. It is the combination or network of relations in which they are placed in the experiment that determines the outcome and interpretation of durability. The purpose of the tests was to determine durability as defined and interpreted by the experienced investigator, not to derive a single measure of durability. In this context the crystallization test fulfilled its function, although the dominance of Portland Stone as the basis for comparison restricted the type of decay viewed as satisfactory within the test. When a statistical assessment was applied to the data a similar result to those obtained by contemporary authors was obtained. Given the construction of the test procedures this is not surprising. Of greater significance is that the BRS data from series 48 refers to weathered Portland Stone, not fresh samples. The relationship between the physical properties of porosity and saturation coefficient are maintained despite varying time and degrees of weathering. There is, however, a covariation between the state of samples and the building from which they were extracted. This could imply that accurate prediction of weight loss from solely stone properties is not feasible except at a very broad level. Moh'd *et al.* suggested that this is also the case for fresh stone samples.[41] In their study, although very 'resistant' stone and very 'weak' stone could be accurately allocated to durability classes A and F, allocation of stone in classes B–D was more difficult.

The above suggests that whilst the crystallization test can be viewed as a construct based on a specific set of entities forming a unique network, some general conclusions can be derived from it and applied to other networks within which the same experimental procedures are used. The most important seems to be that the decay produced reflects the properties of the stone measured, although the relationship between the two is not simple. The nature of individual samples can affect the exact values produced upon running the crystallization test. This means that only broad categories can be assigned to how Portland Stone will respond to the crystallization test, as the current test does in practice.[42] The crystallization test can provide very general limits or indicators of the possible behaviour of stone, it cannot predict how a specific stone will weather in a building. How crystallization relates to durability as a complex phenomenon defined by contemporary authors is a further complication.

Acknowledgements

The authors would like to thank Dr. Tim Yates of the BRE, for permission to dig amongst the files in the archive and also the various BRS personnel from the 1920s and 1930s who kept such good notes and filed them so well that it was possible to interpret their experimental work after 70 years. The authors would also like to thank the constructive comments of two anonymous referees.

References

1. Ross, K.D. and Butlin, R.N. (1989) *Durability Tests for Building Stone*. BRE Report 141.
2. Sedman, J.H.F. and Stanley, L. (1990) Variations in the physical properties of porous building limestones. *Stone Industries* 25: 22–24
3. Sedman, J.H.F. and Stanley, L. (1990b) The crystallization test as a measure of durability. *Stone Industries* 25: 26–28.

4   Ross, K.D. and Massey, S. (1990) A response to the Sedman and Stanley analysis. *Stone Industries* 25: 30–33.
5   B.R.E. (1994) *Standardisation in support of European legislation: what does it mean for the UK construction industry.* BRE Digest 397, September 1994, Ci/SfB(A).
6   B.R.E. (1995) *A guide to attestation of conformity under the Construction Products Directive.* BRE Digest 408, August 1995, Ci/SfB(A).
7   Goudie, A.S. (1999) Experimental salt weathering of limestones in relation to rock properties. *Earth Surface Processes and Landforms* 24: 715–724.
8   Moh'd, B.K., Howarth, R.J. and Bland, C.H. (1996) Rapid prediction of Building Research Establishment limestone durability class from porosity and saturation. *Quarterly Journal of Engineering Geology* 29: 285–297.
9   Turkington, A.V. (1996) Stone durability. In Smith, B.J. and Warke, P.A. (eds.) *Processes of Urban Stone Decay.* Donhead, London, pp. 19–31.
10   Ibid.
11   B.R.E. (1984) *Decay and Conservation of Stone Masonry.* BRE Digest 177.
12   Anon. (1983) *Standard definition of Terms Relating to Natural Building Stone.* American Society for Testing Materials, Philadelphia, Standard C119–83c.
13   Carr *et al.* (1996)
14   B.R.E. (1997) *Selecting Natural Building Stone.* BRE Digest 420, January 1997, Ci/SfBe.
15   Anon., *op. cit.* (1983)
16   Carr, D.D., Strickland, J., McDonald, W.H. and Bortz, S. (1996) Review of durability testing of building stone with annotated bibliography. *Technical Note, American Society for Testing and Materials*, Philadephia: 324–328.
17   Goudie, *op. cit.* (1999)
18   Garden, G.K. (1980) Design determines durability. In Sereda, P.J. and Litvan, G.G. (eds.) *Durability of Building Materials and Components.* ASTM STP 691: 31–37.
19   Nireki, T. (1980) Examination of durability test methods for building materials based on performance evaluation. In Sereda, P.J. and Litvan, G.G. (eds.) *Durability of Building Materials and Components.* ASTM STP 691: 119–130.
20   Duffy, A.P. and OBrien, P.F. (1996) A basis for the evaluating the durability of new building stone. In Smith, B.J. and Warke, P.A. (eds.) *Processes of Urban Stone Decay*, Donhead, London: 253–260.
21   Schaffer, R.J. (1932) *The Weathering of Natural Building Stones.* Department of Scientific and Industrial Research, Special Report 18. HMSO, London. (facsimile reprint Donhead Publishing 2004).
22   BRE, *op. cit.* (1984)
23   BRE, *op. cit.* (1997)
24   Holmes, S. and Wingate, M. (1997) *Building With Lime.* Intermediate Technologies Publications, London.
25   Inkpen, R.J. (1999) Atmospheric pollution and stone degradation in nineteenth century Exeter, *Environment and History*, 5, 209–220.

26 Inkpen, R.J. (2002, in press). Damage to whole structures, patterns of damage. In Brimblecombe, P. (ed.) *The Effects of Air Pollution on the Built Environment: Air Pollution Review, Vol. IV*, Imperial College Press, London, Chapter 14.
27 Lewry, A.J. and Crewdson, L.F.E. (1994) Approaches to testing the durability of materials in the construction and maintenance of buildings. *Construction and Building Materials* 8: 211–222.
28 Barry, A. (1993) The history of measurement and the engineers of space. *British Journal of the History of Science* 26: 459–468.
29 Turkington, *op. cit.* (1996)
30 Ross and Butlin, (1989)
31 Hacking, I. (1983) *Representing and Intervening: Introductory topics in the philosophy of natural science.* Cambridge University Press, Cambridge.
32 Latour, B. and Woolgar, S. (1979) *Laboratory Life: The Social Construction of Scientific Facts.* Sage, Beverly Hills and London.
33 Latour, B. (1987) *Science in Action: How to Follow Scientists and Engineers Through Society.* Open University Press, Milton Keynes.
34 Callon, M. (1991) Techno-economic networks and irreversibility. In Law, J. (ed.) *A Sociology of Monsters: Essays on Power, Technology and Domination.* Routledge, London: 132–161.
35 Law, J. (1992) Notes on the Theory of the Actor-Network: Ordering, Strategy and Heterogeneity. *Systems Practice* 5: 379–393.
36 Demeritt, D. (1996) Social theory and the reconstruction of science and geography. *Transactions of the Institute of British Geographers NS* 21: 484–503.
37 Moh'd *et al., op. cit.* (1996).
38 Brard (as reported by de Thury, H.) (1828) On the method proposed by Mr Brard for the immediate detection of stones unable to resist the action of frost. *Annales de Chimie et de Physique* 38: 160–192.
39 Ibid.
40 Orton, 1911, in *Proceedings American Society for Testing Materials*, volume 19 (only single page reference of p. 268 given in the archive)
41 Moh'd *et al., op. cit.* (1996)
42 Ross and Butlin, *op. cit.* (1989)

4 The Use of Image Analysis for Quantitative Monitoring of Stone Alteration

V. LEBRUN, C. TOUSSAINT and
E. PIRARD

ABSTRACT

This paper presents a methodology to quantify through image analysis the colour alteration of dimension stones caused by weathering. Specific image acquisition techniques and calibration methods are developed including practical solutions to avoid spatio-temporal noise and drift, colour calibration and positioning calibration. The procedure consists of grabbing and comparing images of polished granite tiles before and after accelerated ageing tests. This method is non-destructive, allowing initial and final image acquisitions on the same sample after repositioning with an accuracy of 0.4 millimetres. Lighting properties and grabbing parameters are kept identical during both imaging phases. Colour alteration is quantified by computing the Euclidean distance in a (pseudo)-L*a*b* colour space. The results of a practical study on three granites are presented and discussed.

INTRODUCTION

The aim of this paper is to propose technical solutions and analytic methods for monitoring the colour decay caused by weathering using digital image analysis. The mineralogy and sampling of the three studied granites are first briefly described, followed by an overview of the complete protocol of tests. The two main sections develop the calibration procedure for the imaging system and the image analysis methods. Finally, the most significant results are presented and commented upon, with a brief discussion of potential extensions of the proposed techniques and its limitations.

The quantitative description of visual change in ornamental stones submitted to natural or artificial weathering is a crucial scientific and economic challenge. Colour is one of the most important characteristics of these materials, which determines their use in the building market. Mineralogical or geochemical mechanisms of rock alteration have been widely studied.[1] Some studies have tried to quantify the degradation of mechanical and physical properties,[2] but very little information is available about the aesthetic alteration of stones after weathering. However, from the consumer's point of view, visual appearance of stone is probably the most important consideration. For example, any deviation, even slight, in the predominant colour of adjacent tiles in a paving badly affects the aesthetics of the work as a whole by emphasizing the colour discontinuity induced by the cement joints.

Of course, standard colorimeters or spectrophotometers are available for quantitative measurement of colour. But both devices integrate very limited fields of investigation (typically 1 cm^2). They are thus largely irrelevant for monitoring colour variations in textured materials like granites.

The idea of using digital image analysis to characterize or control the quality of ornamental stones or ceramic is not a revolutionary idea.[3] For the last five years various consortia have managed ambitious research programs at the European level with the aim of integrating automatic inspection systems for ceramic tiles.[4,5] This recent interest in quantitative colour quality control using digital video cameras demonstrates the usefulness and further potential of the technique and may meet the need for an accurate and objective tool to control the visual aspect of tiles in the stone industry.

The work described in this paper extends these studies by conducting colour image analysis over a longer time scale, which is relevant to colour alteration by weathering and colour durability. The test methods used have been applied under laboratory conditions, but could be adapted, with some limitations, for use *in situ* on paving, façades and natural stone outcrops.

MATERIALS AND METHODS

Granite selection

For this study, three granitic rocks were selected (Table 4.1) for their high propensity for chromatic alteration.[6] Three polished flags (300×300 mm) of each rock type were prepared and analysed. The tiles were randomly selected from the daily production of the source quarries.

General testing protocol

To ensure representativeness and reproducibility, it is essential that a rigorous test protocol is devised and strictly adhered to. Each fresh sample tile is first imaged and its initial colour statistics are computed and stored. The tiles are then submitted to standardized accelerated ageing tests. After precise repositioning, a second colour image is acquired from which final colour statistics are computed.

Table 4.1 Names, origin and mineralogy of the studied granites.

Rock type	Nature	Origin	Principal	Secondary	Accessory
Tarn	granite	France	Albite Microcline Quartz Biotite		Kaolinite Chlorite
Azul Paveno	gneiss	Brazil	Albite Microcline Quartz	Biotite	Kaolinite
Baltic Brown	Wiborgite (granite)	Finland	Albite Microcline Quartz	Biotite Hornblende	

Finally, the colour alteration is computed by combining 'before' and 'after' colour data. Figure 4.1 details the flow-sheet of the protocol.

Accelerated ageing tests

The ageing chamber used is a KSE-300 designed for the realization of corrosion tests following the DIN50017 and DIN 50018 norms and for alternating condensation controls.[7,8]

The following standardized tests were performed:

♦ acid test (2 samples of each stone type): 21 days at 20°C in a SO2 saturated atmosphere;
♦ water and heat test (1 sample of each stone type): 21 heating cycles during 20 hours in a ventilated drying-room at 105°C followed by 4 hours in a water bath at 20°C.

These tests have been chosen for two main reasons. Firstly, Elsen observed some visually significant colour changes in the same stone submitted to the acid test;[9] secondly, both tests use oxidizing atmospheres suitable for modelling the urban environment to which external wall tiles could be exposed during their life.

THE USE OF IMAGE ANALYSIS 57

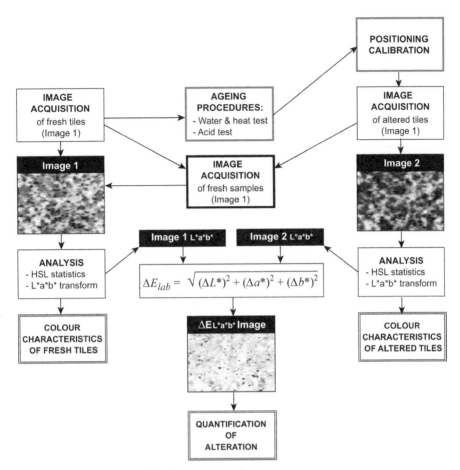

Figure 4.1 Flow-sheet of analyses protocol.

Imaging devices

A specific acquisition system was designed to take into account the optical properties of the samples. The three main characteristics of the polished granite tiles are:

- macroscopic dimensions (300×300 mm)
- high reflectivity
- textured colour

These properties induce technical constraints on the imaging system,[10] which needs to have the following characteristics:

- diffusing
- 'daylight' white
- homogenous
- stable in time

In practice, the final solution adopted for this study (Figure 4.2) is a wide imaging box equipped with high frequency 'daylight' luminescent tubes, diffusing walls and ceiling and a black and white CCD camera (1280×1024 pixels) with red green and blue gelatine colour filters.

CALIBRATION OF THE IMAGING SYSTEM

As for any scientific instrumentation, each step of the image acquisition with CCD-camera must be controlled and calibrated. The combination of lighting, optics and electronic defaults induces high frequency noise and low frequency drift in time and space. Moreover, each optical and electronic component of the system acts as a colour filter, contributing to a global colour deviation.

Correction for temporal noise and drift

Low frequency temporal drift, due to the pre-heating of the CCD and to the ageing of each imaging device, is avoided by switching on the camera at least two hours before the acquisition and by frequent calibrations of the imaging system. The high frequency variations of the signal through time can be decomposed into electronic noise and thermal noise.[11] For the camera used in this study this combined noise reaches 5 grey levels of amplitude. It is filtered by averaging a

THE USE OF IMAGE ANALYSIS 59

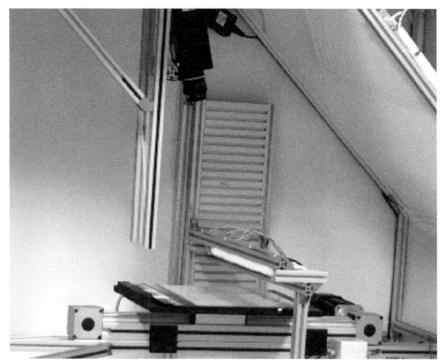

Figure 4.2 The imaging system.

temporal sequence of 16 images for each analysed scene. Wider averaging does not further improve the quality of the acquisition.

Correction for spatial noise and drift

Non-uniform illumination and optical aberrations cause a low frequency drift across the field of view. Even when it does not receive any photons, each cell of the CCD produces a specific non-null response (dark current). This results in a high frequency spatial noise called 'black noise'.[12] By tuning the gain and offset of each individual pixel,[13] one can compensate for both spatial drift and black noise. This dual operation is detailed below.

The 'white noise' image, where intensity is proportional to the drift, is grabbed with a glossy white PEHD (high density polyethylene) plate giving a good compromise between whiteness and reflectivity. The 'black noise' image is obtained by closing the diaphragm of the camera. Dividing each acquired picture by the white image allows balancing of the low frequency spatial drift. Subtracting the black image from the result achieves the black noise correction.

In the case of an 8 bits signal coming from a single colour channel, for example, red, the corrected image C^R is obtained, from the input image I^R, by the following relationship:

$$C^{R'}_{(x,y)} = \left[\frac{\left(I^R_{(x,y)} - Bl^R_{(x,y)}\right)}{\left(Wh^R_{(x,y)} - Bl^R_{(x,y)}\right)} \times 255 \right] \quad (1)$$

where:

i. $I^R_{(x,y)}$ is the pixel of co-ordinates (x,y) in the input image I, red channel;

ii. $Bl^R_{(x,y)}$ is the same pixel in the black noise image, red channel.

iii. $Wh^R_{(x,y)}$ is the same pixel in the white image, red channel.

iv. 255 is the maximal value for a 8 bit digital signal.

The spatial and temporal correction procedures are illustrated in Figure 4.3. The first image shows the white sample without correction. The second image shows the application of a look-up table in order to sharpen the low frequency drift. This only gives a better visualization of the phenomenon; images 1 and 2 are therefore exactly the same. Images 3 and 4 show the results respectively after temporal averaging and background correction. The grey level profiles traced across the images clearly show the effect of each correction. The second profile is less noisy (temporal noise erased) than the first; the third profile does not present any curvature (spatial drift) and it is smoother than the second profile (black noise erased). The remaining 'noise' in the corrected image is due to the texture of the

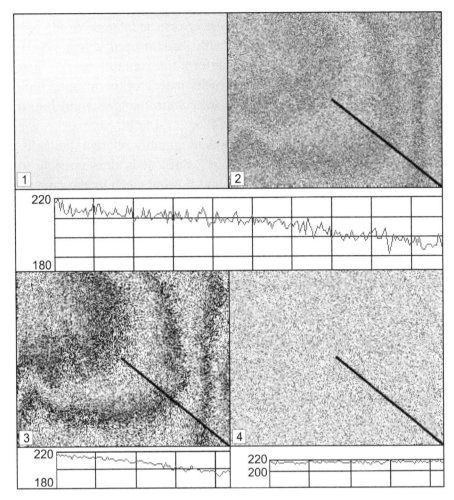

Figure 4.3 Spatio-temporal drift correction.

reference sample itself. This problem can be solved by application of an alternating sequential low-pass filter on the resultant image.[14]

Colour calibration

Each optical component of the acquisition device induces deviation of the output red green and blue channel compared to the intrinsic colour of the observed material. These errors can be numerically corrected by calibrating the system with standardized colour charts. Various solutions are available to perform this calibration.[15,16,17] Their common goal is to compute the transfer-matrix of the imaging tool, which converts the device-dependant colour channels (rgb) into a calibrated colour system like CIE-RGB or CIE-L*a*b*.

In this study, the main objective is to quantify relative shade differences rather than absolute colour values. It is thus possible to work directly with the output R, G and B value given by the camera. Nevertheless, an original procedure is used to optimize the colour rendering of the system to ensure its reliability. Each image is equipped with a standard grey scale (Kodak Q-14). The integration time, the gain and the offset of each channel are tuned individually in order to obtain identical mean intensities in the grey patches of the scale. Although the obtained colour co-ordinates remain device-dependant, this calibration ensures the reliability of the relative colour deviation measure, as the same device is used for both 'before' and 'after' acquisition phases.

In future, it would be interesting to improve the method by computing the transfer-matrix of the system using a large reference colour set.[18] The colour set should be chosen in order to cover the entire shade gamut of each rock type under investigation. This would allow standardization of the measure of colour alteration in the CIE-L*a*b* system.

POSITIONING CALIBRATION

Image analysis has the great advantage that it is non-destructive. In order to exploit this property, one has to take pictures of the same sample before and after accelerated ageing and compare these two images. For precise repositioning to be possible, it is necessary to fix three spatial references on each tile before starting the image acquisition procedure (Figure 4.4). These spatio-referenced targets allow images of fresh and altered tiles to be combined with a spatial

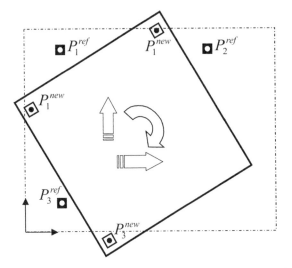

Figure 4.4 Positioning Calibration: the tile is moved until the new co-ordinates of the targets P_i^{new} are identical to the spatial references P_i^{ref} stored at the initial image acquisition.

accuracy of a single pixel (0.4 mm). The image grabbing geometry is kept identical during both acquisition phases (the system is locked during the ageing tests). The positioning operation is thus not affected by lens distortion, which is equal in both 'before' and 'after' images.

IMAGE ANALYSIS

HSL global statistics

Global colour deviations are measured on entire tiles in the hue-saturation-intensity space, HIS. This is a bi-conical colour space, directly deduced from RGB by linear transforms (Figure 4.5). Of course, it is non-linear and 'device-dependent', but it presents the great advantage that it is probably the most intuitive manner of specifying colour. The relative 'before-after' differences between mean global values of each channel are computed as follows:

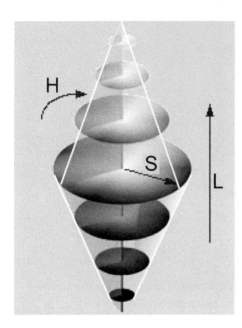

Figure 4.5 The HSL colour space (Russ, J.C. (1999) *The Image Processing Handbook*. 3rd ed. CRC press, Boca Raton. p. 50).

$$dH = \frac{2 \bullet (\overline{H}_{out} - \overline{H}_{in})}{\overline{H}_{out} + \overline{H}_{in}} \quad dS = \frac{2 \bullet (\overline{S}_{out} - \overline{S}_{in})}{\overline{S}_{out} + \overline{S}_{in}} \quad dL = \frac{2 \bullet (\overline{L}_{out} - \overline{L}_{in})}{\overline{L}_{out} + \overline{L}_{in}} \quad (2)$$

where $\overline{H}_{out}$ is the mean of hue channel in output image

These values are convenient for explaining in words the visual observations of global colour deviations, but they must not be considered as quantitative measure of these changes. Even Euclidean distance in HSI space would not give such an objective quantification, as it is not a linear space with regard to visual perception.[19] Nevertheless, for the industrial transfer of the methodology, the utilization of self-understanding colour decomposition may be necessary.

L*a*b* local distances

In order to map the quantitative colour variation undergone by each pixel during the accelerated ageing operation, it is necessary to use a perceptually linear space like CIE L*a*b*. This is a uniform spherical colour space especially created by the CIE for colour difference computations (Figure 4.6). The transformation from R, G, B channels to L*a*b* is non-linear. For the complete description of the transfer equations, the reader can refer to the CIE recommendations.[20]

These relations are only relevant for calibrated CIE-RGB values. Applying the same formula to the device-dependant red green and blue channels would make no sense. In practice, the obtained pseudo L*a*b* system has been preferred because it will allow future work with calibrated CIE-RGB inputs, using the same method for CIE-L*a*b* computation.

Once the L*a*b* co-ordinates are computed, the difference between two measured colours is given by the following relation:

$$\Delta E_{Lab} = \sqrt{(\Delta L^*)^2 + (\Delta a^*)^2 + (\Delta b^*)^2} \qquad (3)$$

In the present application, ΔL^*, Δa^* and Δb^* are the pixel-to-pixel differences within each of the L*, a* and b* channels before and after the accelerated ageing tests. The resulting distance function is mapped in a new image in which the intensity of each pixel is proportional to the computed L*a*b* colour distance at the corresponding point of the tile.

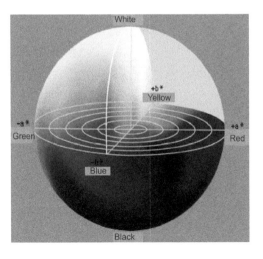

Figure 4.6 The CIE-L*a*b* colour space (Russ, J.C. (1999) *The Image Processing Handbook*. 3rd ed. CRC press, Boca Raton. p. 50).

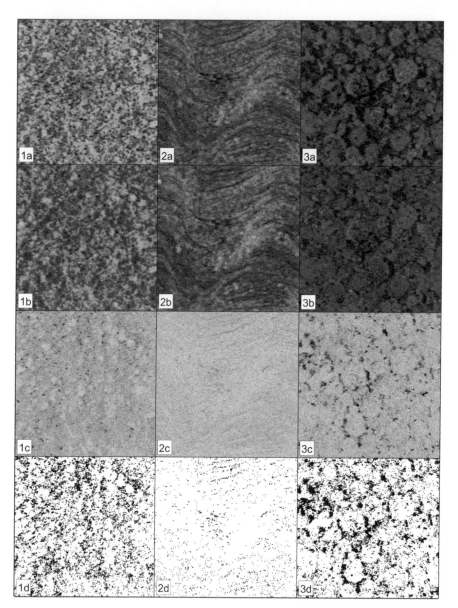

Figure 4.7 Images of each stone type studied – Top to bottom of figure: a) fresh samples; b) altered samples; c) distance map. Left to right of figure: 1) Tarn; 2) Azul Paveno; 3) Baltic Brown.

The mean and maximum values of the distance map give a more quantitative, but less intuitive, idea of the global colour deviation than the HSL statistics.

By choosing an appropriate threshold in the distance map (Figure 4.7), it is possible to extract the most altered zones in the tiles and to compute some useful statistics for them including:

- number of zones,
- relative cumulated area of the zones,
- mean individual size of zones
- mean colour distance computed on the zones only.

Figure 4.7 shows the input images of original and altered stone, the corresponding distance-map and the binary images resulting from the thresholding operation.

RESULTS

Macroscopic observations and mineralogy

The 'water and heat' test did not cause any perceptive chromatic alteration. No significant colour change can be visually detected. However, the parameters of this test (temperature, moisture, O_2 and CO_2) are expected to favour the oxidation of pyrite into limonite or goethite.[21] The absence of this phenomenon can be due to a very low pyrite content or to very slow oxidation.

In comparison, some of the samples exposed to the acid test locally underwent drastic colour changes especially those of Tarn and Baltic Brown.

Yellowish white efflorescences appeared on the three studied stones. These crusts, identified by optical microscopy, are composed of gypsum ($CaSO_4.2H_2O$) and are more abundant Baltic Brown granite. Elsen explains this relative abundance by the presence, in the later, of calcic hornblende, which probably catalyses the crystallization.[22] He argues that this mineral contains Ca^{++} ions needed for gypsum formation and does not take a good polish, thus offering a larger specific surface for alteration.

Rust coloured patches (limonite and goethite) appear on Tarn and, to a lesser extent, on Baltic brown tiles. In both cases, careful macroscopic observation allows the identification of muscovite as the preferential occurrence site.

Finally, a global tarnishing of the colours is observed on every stone type. This apparent tarnish is due to the degradation of the surface state, which induces a decrease in reflectance. The surface state alteration is visually more intense in Tarn than in Baltic Brown granite. The Azul Paveno remains almost unaltered.

HSL statistics

Table 4.2 contains the relative HSL differences between fresh and altered samples (dH, dS and dL). These differences are computed by equation (1).

L*a*b* distances

The most significant global and local parameters computed from the distance map images are presented in Table 4.3. The first two columns contain the mean and maximum distances computed by equation (2) on the full surface of the tiles. Columns 3 to 6 depict local information about the most altered zones with, from left to right: the total number of zones, their cumulated area proportion, their mean individual size and the mean colour distance computed on them.

Table 4.2 HSL relative differences between altered and fresh samples.

Rock Type	dH (%)	dS (%)	dL (%)
Tarn	-33	+74	-28
Azul Paveno	-12	-27	+2
Baltic Brown	-26	+37	-10

Table 4.3 Global and local analysis of the CIE-L*a*b* colour distance images.

Rock Type	Global statistics		Statistics on the most altered zones (colour distance >13)			
	Mean dist.	Max. dist.	Nb. zones	Area (%)	Mean size	Mean dist. in zones
Tarn	10.4	66.8	3282	21.4	8.3	15.3
Azul Paveno	7.0	35.2	2858	4.1	1.9	15.1
Baltic Brown	10.5	74.4	3400	23.4	8.9	15.6

DISCUSSION

The HSI differences presented in Table 4.2 are not really representative of colour changes, mainly because they give global information when colour alteration is often local. Nevertheless, they allow the expression of visual observations in numerical terms and confirmation that the digital imaging system is able to model human eye perception, at least up to a certain level.

Table 4.3 confirms that the most important chromatic variations are observed in the Tarn and Baltic Brown granites. The intensity channel is affected by two opposite interactions: the surface state alteration tends to decrease the intensity, whereas the white efflorescence brings the intensity up. By comparing the three stones in column 3 of Table 4.2, one can deduce that the tarnish effect is predominant in Tarn and, to a lesser extent, in Baltic Brown. For Azul Paveno, surface state and efflorescence effects compensate for each other. The sharp increase in saturation observed for Tarn and Baltic Brown is probably due to rust coloured patches having high saturation levels compared with the duller shades of natural granites.

The L*a*b* distances maps and their related parameters give a much more precise description of the intensity of colour deviations. The information contained in Table 4.3 can be summarized as follows:

- The mean colour distances are in complete concordance with the HSI differences concerning the global alteration of each stone type.
- The lower global colour deviation observed in granite Azul Paveno is not due to lower mean colour distances in the altered sites. It results from a lower number and a smaller mean size, thus a lower surface proportion, of these zones.
- Although, Baltic Brown and Tarn present similar global colour alterations, the spatial distribution of this alteration is somewhat different. The Tarn granite contains fewer patches, with lower mean size and lower mean distance than Baltic Brown. This could result in a higher global mean distance in the latter.

SCOPE FOR APPLICATION AND LIMITATIONS

Weathering of stones is a subject touching many fields of study. Geomorphologists trying to understand the mechanism of landscapes evolution, archaeologists and architects working on the restoration of ancient heritage; material engineers comparing the durability of building stones; quality engineers of quarries involved in certification and technical agreement policies. All of them tackle weathering problems from their own point of view, technical background, and with their own investigation tools. The tools presented in this paper are of course not universal, their approach locates them on the frontier between material science, electronics and computer science. It could be argued that these sciences are far from, for example, geomorphology's usual preoccupation. Digital imaging is a very promising technique, however, with a wide perspective of application within many disciplines.

Digital image analysis is not destructive. In comparison with other tools used for durability assessment like mechanical tests or chemical analysis, measurements are taken on the same sample before and after alteration. This study has not examined the relationships between accelerated ageing tests and the reality of stone weathering and alteration. It is assumed that these tests do, however, provide useful information about the alteration of material properties by 'real' weathering processes. What can be said, is that in this case, calibrated digital image analysis provides a meaningful, quantitative and reliable evaluation of colour deviation.

Colour is not the only physical property that can be monitored by digital imaging. Under oblique light, one can take pictures in which the intensity is inversely proportional to the roughness of the surface. Image analysis could therefore be used, with similar *modus operandi*, to quantify the loss of lustre induced by weathering on polished stones.

If the exact positions of altered zones are known; information about them could be used in a geographical information system, to relate alteration to factors such as mineralogy, chemistry, microhardness or surface state.

Digital imaging is not confined to laboratory experimentation; a video camera might be used to observe wall cover material in place on building façades. Real scale studies can also be managed by grabbing photographs of façades and comparing them with pictures taken subsequently. One final application of image analysis is the study of colour change and alteration in natural outcrops.[23]

The main limitations of imaging techniques are linked to image acquisition reliability. If it is not possible to take photographs before and after weathering in exactly the same conditions (geometry, field of view, lighting) colour and positioning calibrations are much more critical. In any case, taking time to optimize image acquisition is never time lost. The information lost by using an inadequate or uncalibrated imaging system is lost forever.

The topography of observed surfaces can induce other problems. Optical aberrations and geometric deformations and colour shading occur with irregular surfaces. Numerical corrections exist, which can attenuate these problems but never eliminate them.

CONCLUSIONS

The results presented above, show that digital cameras and image analysis can perform quantitative colour measurements of large textured samples. Combined with accelerated ageing tests, image analysis is a powerful tool for the evaluation of colour deviations caused by weathering on ornamental stones. But the most important point to remember is the huge potential of digital image analysis as an investigative and diagnostic tool for earth and material scientists.

References

1. Delvigne, J. (1998) *Atlas of Micromorphology of Mineral Alteration and Weathering*. Mineralogical association of Canada, Ottawa.
2. De Cleene, M. (1995) Interactive physical weathering and bioreceptivity study on building stones, monitored by computerized x-ray tomography (CT) as a potential non-destructive research tool. *Commission of the European communities, directorate-general for science*, research and development, Bruxelles.
3. Baldrich, R., Vanrell, M. and Villanueva, J.J. (1999) Texture-colour features for tile classification. *EUROPTO/SPIE Conference on Colour and Polarisation Techniques in Industrial Inspection*, Munich, Germany.

THE USE OF IMAGE ANALYSIS 73

4   Maccari, A. (1998) ASPECT: an intelligent sorting system. *Ceramic World Review* 8: 138–141.
5   Donia, G. (2000) Inspector 2000 project. *Ceramic World Review*, 10: 216–219.
6   Elsen, J. (1996) *Diagnostic des pierres naturelles et optimisation de leurs techniques de mise en œuvre, T.2 : Le comportement des granits sous l'influence de l'exposition en atmosphère acide.* Centre scientifique et technique de la construction, Bruxelles.
7   DIN 50017-Ausgabe:1982–10 (1982) *Klimate und ihre technische Anwendung; Kondenswasser-Prüfklimate.* Deutsches Institut für Normung e.V. Berlin.
8   DIN 50018-Ausgabe:1997–06 (1997) *Prüfung im Kondenswasser-Wechselklima mit schwefeldioxidhaltiger Atmosphäre.* Deutsches Institut für Normung e.V. Berlin.
9   Elsen, *op. cit.* (1996)
10 Lebrun, V., Bonino, E., Nivart, J.-F. and Pirard, E. (1999) Development of specific acquisition techniques for field imaging-Applications to outcrops and marbles. In *Proceedings of the International Symposium on Imaging Applications in Geology*: 165–168.
11 Holst, G.C. (1998) *CCD Arrays, Cameras and Displays*, SPIE Optical Engineering Press, Washington: 127–131.
12 Holst, *op. cit.* (1998)
13 Pirard, E. Lebrun, V. and Nivart, J.-F. (1999) Optimal acquisition of video images in reflected light microscopy. *Microscopy and Analysis* 60: 9–11.
14 Soille, P. (2000) *Morphological Image Analysis.* Springer, Berlin.
15 Connoly, C. and Leung, T. (1995) Industrial colour inspection by video camera. *IEE Int. Conf. On Image Processing and its Applications*, Conference publication No.410, IEE: 672–676.
16 Chang, Y.-C. and Reid, J. (1996) RGB calibration for color image analysis in machine vision. *IEEE Trans. Image Proc.* 5: 1414–1422.
17 Marszalec, E. and Pietikäinen M. (1997) Color measurements based on a color camera. In Ahlers, R.J. and Réfrégier, P. (eds) New image processing techniques and applications: algorithms, methods and components II, Proceedings of SPIE, June 1997: 170–181.
18 Marszalec, E. and Pietikäinen M. (1996) Some aspect of RGB vision and its applications in industry. *International Journal of Pattern Recognition and Artificial Intelligence* 10: 55–72.
19 MacAdam, D.L. (1942) Visual sensitivities to color differences in daylight. *Journal of Optical Society of America* 32: 247–273.
20 CIE. (1986) Colorimetry. Official recommendations of the International Commission on Illumination, CIE Publication No. 15.2, Vienna.
21 Costagliola, P., Cipriani, C. and Manganelli, F.C. (1997) Pyrite oxidation: protection using synthetic resins. *European Journal of Mineralogy*, 9: 167–174.
22 Elsen, J. (1998) *Durability of Granites for Construction.* Aardk. Mededel., 9: 35–40.
23 Lebrun *et al, op. cit.* (1999)

5 Mechanisms of Attack on Limestone by NO_2 and SO_2

G.C. ALLEN, A. EL-TURKI, K.R. HALLAM,
E.E. COULSON and R.A. STOWELL

ABSTRACT

This paper provides details on chemical mechanisms involved in the attack upon calcareous stones by oxides of nitrogen and sulphur. Samples of Bath and Portland Stone, as-cut and after application of proprietary siloxane- and fluorocarbon-based water-repellent surface treatments, were exposed to sulphur dioxide, nitrogen dioxide and both gases combined under humid conditions and analysed subsequently using x-ray photoelectron spectroscopy (XPS). The synergistic relationship between the two gaseous pollutants, previously found to accelerate degradation of the stone, especially in the presence of water, is rationalized using a molecular orbital model.

INTRODUCTION

Urban stone decay is a complex phenomenon.[1] For example, on calcareous stones the rate of attack depends on many factors, including relative humidity, but is greatest for SO_2 in the presence of NO_2 or ozone. This interplay of reactants is especially important in city environments. To this end, laboratory rigs have been designed, constructed and commissioned to simulate, in a controlled way, the effect of single and mixed pollutants on stone from specific sources. The stone has been characterized before and after exposure and individual samples have been exposed to various environments involving NO_2, SO_2 and constant humidity.

A strong synergistic effect on stone degradation has been reported for NO_2 and SO_2 mixtures in ppm and sub-ppm ranges in humid air,[2,3,4] with NO_2 said to act as a catalyst for the oxidation of SO_2 on the stone surface. Here we report changes identified using x-ray photoelectron spectroscopy (XPS) as an early indication of the reaction taking place on limestone surfaces and provide further details of the mechanism of this reaction.

There has also been much interest in the use of resin-based coatings to protect both old and new stone buildings. Such treatments are designed to minimize water and oil penetration, dirt accumulation and the growth of lichen and moss, to be long lasting, invisible, UV-resistant and yet to allow the natural material beneath to 'breathe'. Their water resistant properties are thought to allow rain to form compact drops and to assist in the removal of particulates from the surface. With this in mind, we have investigated the effects of four proprietary treatments following exposure to NO_2 and SO_2 pollutant atmospheres.

EXPERIMENTAL PROCEDURES

The use of especially sensitive techniques, such as x-ray photoelectron spectroscopy, has enabled data to be acquired from surface reactions at an atomic level. Thus, the progress of stone decay may be detected much sooner, and in greater detail, than would be possible by other, more conventional, means. We have investigated Bath Stone (Monk's Park) and Portland Stone, as representative of stones widely used in building construction in the UK. Small samples measuring approximately 10×10×5 mm were cut from large slabs using a circular diamond saw under dry conditions. The conditions for the controlled exposure experiments undertaken on the Bath Stone have been described previously.[5,6,7]

Following exposure, the surfaces were analysed using a ThermoVG Scientific Escascope x-ray photoelectron spectrometer with a magnesium K_α (1253.6 eV) source, operating at 300W (20mA, 15kV). Spectra were recorded from an area of $c.$4×3 mm over the entire energy range before concentrating on the carbon $1s$, nitrogen $1s$, oxygen $1s$, silicon $2p$, sulphur $2p$ and calcium $2p$, photoelectron regions. All spectra were corrected for charging effects with respect to the adventitious hydrocarbon carbon $1s$ peak at 284.8 eV binding energy and quantified using the VGS5250 operating software sensitivity factors.

To investigate the effects on stone corrosion of surface treatments, coated samples of Bath and Portland Stone were similarly exposed to humid NO_2, SO_2 and NO_2+SO_2 gases. Labelled here as treatments A, B, C and D, the coatings were, respectively, a fluorocarbon resin based coating, a mixed fluorocarbon and siloxane based coating, a fluorocarbon based coating notably containing phosphates and a second siloxane and fluorocarbon coating. These were applied to one surface of the individual samples using a brush, in accordance with the manufacturers' instructions. Energy dispersive x-ray and x-ray photoelectron spectra were recorded from the stones as-received, and after coating but before exposure, to provide baselines for subsequent analysis. The samples were housed in a glass vessel such that the coated faces to be analysed afterwards were facing the oncoming gas flow. Temperature and humidity monitoring, gas mixing and humidifying were achieved as reported earlier.[8,9,10]

After exposure, the samples were analysed using x-ray photoelectron spectroscopy, with particular notice being paid to the extent of sulphation of the sample surface and the degradation of the coatings.

RESULTS

Results of the initial pollutant gas exposures of uncoated Bath Stone have been presented previously.[11,12,13] Only a summary of the main findings will be given here.

Samples exposed to dry NO_2 showed the presence of the reaction product calcium nitrate together with a second species identified as adsorbed NO. The latter had previously been observed only on hydrated silica and alumina-silica surfaces exposed under similar conditions.[14] The presence of NO was not reported in the outflow gases of Johansson's experiments[15] but, at high relative humidity, NO is very rapidly oxidized to NO_2. For samples exposed to wet NO_2, however, calcium nitrate was the predominant product. Although the amount of adsorbed NO was similar under both dry and humid conditions, the extent of nitrate was much enhanced in the latter due to the ease of formation of nitric acid when NO_2 is in the presence of water vapour. The results obtained in the presence of both NO_2 and SO_2 are of special interest. Under both wet and dry conditions, calcium sulphate was the only sulphur-containing phase produced, but the extent of the reaction was some ten times greater under the former.

These results show that NO_2 increases the rate of attack by SO_2, but does not strictly play a catalytic role. The presence of NO_2 accelerates the rate of sulphation of limestone by aiding the oxidation of sulphur dioxide to sulphur trioxide, usually a slow process. In turn, sulphur trioxide reacts with the calcareous stone to form calcium sulphate. The reaction proceeds, in the presence of traces of sulphuric acid, via the formation of nitrosonium hydrogen sulphate. NO_2 is initially consumed, forming NO, which may then be readily oxidized back to NO_2. The presence of small amounts of sulphuric acid also increases the rate of NO formation.

The presence of NO on the surface of the reacted stone is a significant finding. Below we explain in detail how it is bonded to the calcite and is able to become available for oxidation back to NO_2 and to facilitate further sulphation of the stone.

XPS analysis of coated stone prior to gas exposure revealed that the application of the coatings had not completely covered the surface. XPS analyses to a depth of only $c.5$ nm. However, calcium was still detected in all analyses (both stones and all four coatings). The extent of the coating was estimated by looking at the percentage fall in the measured calcium surface atomic concentration due to the presence of the coating. Coating C (fluorocarbon/phosphate) was found to provide the most complete coverage and B (siloxane/ fluorocarbon) the least complete, with A (fluorocarbon) and D (fluorocarbon/siloxane) giving intermediate values. This effect would appear to be a function of the viscosities and compositions of the coating solutions and suggests that in all cases $CaCO_3$ would still be available for reaction with the pollutants in the atmosphere. Overall, greater coverage appeared to be achieved on the Portland Stone, which could be explained by its porosity characteristics that present a more uniform surface to the coating, allowing a more complete coat to be prepared, whereas the more porous Bath Stone would allow the coating, while still liquid, to settle into the pores, leaving more calcite exposed.

The effect of this exposed calcite was manifested in the observations of sulphur present on almost all samples after exposure, though generally to a lesser extent on the Portland Stone samples. The least protection, indeed higher concentrations of surface sulphur than were seen on uncoated specimens after exposure, was provided by coating B (siloxane/fluorocarbon), which had been seen before exposure to have covered the limestone surfaces the least. Figure 5.1 shows the nitrogen and sulphur regions for Monk's Park coated in B (siloxane/fluorocarbon) following exposure to NO_2, SO_2 and NO_2+SO_2. As with the earlier study, it was found that the greatest increases in surface sulphur concentration were when both pollutant gases were present. The expected mixture of sulphite and sulphate following exposure to SO_2 alone and pure sulphate after NO_2+SO_2 are clearly seen. Conversely, the greatest protection against sulphation

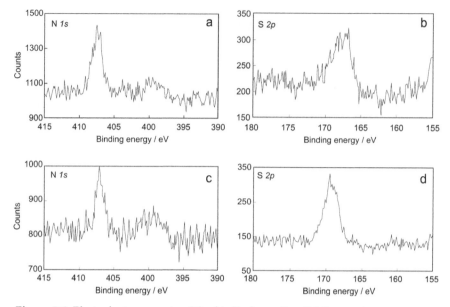

Figure 5.1 Photoelectron spectra, Monk's Park, coating B (siloxane/fluorocarbon): a) N $1s$ after NO_2; (b) S $2p$ after SO_2; (c) N $1s$ after NO_2+SO_2; (d) S $2p$ after NO_2+SO_2.

would appear to have been provided by coating C (fluorocarbon/phosphate), which was seen earlier to have given the greatest apparent coverage of the limestone. Indeed, for the Portland Stone, no sulphation at all was observed, either with SO_2 as the sole pollutant gas or with NO_2+SO_2. Likewise, no surface sulphur was recorded for the Monk's Park specimen following NO_2+SO_2 exposure. Figure 5.2 shows the nitrogen and sulphur regions for Monk's Park coated in C (fluorocarbon/phosphate) following exposure.

Peak fitting was used to identify the relative amounts of calcium sulphite and hydrated calcium sulphate (gypsum) for each sample, with peaks at $c.167$ eV and $c.169$ eV respectively.[16,17] The results are shown in Tables 5.1 and 5.2. Both types of stone, in their uncoated state, showed the presence of both sulphite and sulphate after exposure to SO_2 on its own, whereas when both gases were present, only peaks attributable to the sulphate were observed. Similar results were obtained for the majority of the coated samples, suggesting that they have no qualitative effect on the reaction processes occurring at

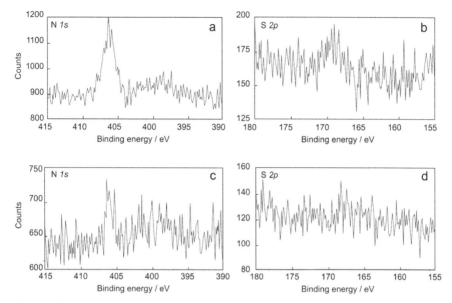

Figure 5.2 Photoelectron spectra, Monk's Park, coating C (fluorocarbon/phosphate): (a) N 1s after NO_2; (b) S 2p after SO_2; (c) N 1s after NO_2+SO_2; (d) S 2p after NO_2+SO_2.

the surface of that portion of the stone surface not covered by coating.

Looking at the results for nitrogen adsorption at the surface, we again see that coating B (siloxane/fluorocarbon) gave the least protection against attack by NO_2, and C (fluorocarbon/phosphate) the greatest. In addition to calcium nitrate and adsorbed NO, peaks representing calcium nitrite were seen on two samples – Monk's Park uncoated and Portland B (siloxane/fluorocarbon), both following exposure to NO_2.

For coating B (siloxane/fluorocarbon) a decrease in the surface concentrations of fluorine and phosphorus was observed under all exposure conditions. At the same time though the values for silicon and calcium increased. It would appear that the fluorocarbon and phosphate in the coating material acted in a sacrificial manner to protect the stone from the pollutants. In being oxidized and lost to the gas stream, more calcite surface was being revealed, which would, of course, become vulnerable to attack from the NO_2 and SO_2. Similar

Table 5.1 Binding energies and intensities (atomic percentages) in the spectra from the Monk's Park samples. Binding energy/eV (concentrations/at %)

Sample		S $2p_{3/2}$			N $1s$		
		Sulphite	Sulphate	Oxide	Nitrite	Nitrate
Monk's Park uncoated	SO_2	166.7 (0.6)	169.2 (1.0)	----	----	----
	NO_2	----	----	399.7 (0.8)	404.0 (0.4)	407.6 (0.9)
	NO_2+SO_2	----	169.0 (3.6)	399.9 (0.9)	----	407.5 (0.4)
Monk's Park A (fluorocarbon)	SO_2	166.5 (0.3)	169.0 (0.5)	----	----	----
	NO_2	----	----	399.3 (1.1)	----	407.1 (1.0)
	NO_2+SO_2	167.0 (0.4)	168.8 (1.4)	399.6 (0.4)	----	406.8 (0.3)
Monk's Park B (siloxane/fluorocarbon)	SO_2	167.2 (2.4)	168.7 (0.7)	----	----	----
	NO_2	----	----	399.4 (1.2)	----	407.5 (2.5)
	NO_2+SO_2	----	168.9 (4.4)	399.6 (2.3)	----	407.1 (1.8)
Monk's Park C (fluorocarbon/phosphate)	SO_2	----	169.1 (0.3)	----	----	----
	NO_2	----	----	399.4 (0.8)	----	406.4 (1.6)
	NO_2+SO_2	----	----	399.8 (0.4)	----	406.1 (0.3)
Monk's Park D (fluorocarbon/siloxane)	SO_2	167.0 (1.6)	169.1 (0.5)	----	----	----
	NO_2	----	----	399.4 (1.3)	----	407.2 (0.8)
	NO_2+SO_2	168.0 (0.6)	169.4 (1.5)	399.6 (3.0)	----	407.4 (0.3)

Table 5.2 Binding energies and intensities (atomic percentages) in the spectra from the Portland samples. Binding energy/eV (concentrations/at %)

Sample		S $2p_{3/2}$		N $1s$		
		Sulphite	Sulphate	Oxide	Nitrite	Nitrate
Portland uncoated	SO_2	167.2 (0.7)	169.4 (1.0)	----	----	----
	NO_2	----	----	399.8 (1.0)	----	407.3 (1.0)
	NO_2+SO_2	----	168.8 (3.4)	399.8 (2.1)	----	407.6 (0.5)
Portland A (fluorocarbon)	SO_2	167.2 (0.4)	169.2 (1.2)	----	----	----
	NO_2	----	----	399.6 (0.8)	----	407.0 (0.4)
	NO_2+SO_2	167.2 (0.9)	168.0 (1.2)	399.1 (1.3)	----	406.9 (0.2)
Portland B (siloxane/fluorocarbon)	SO_2	164.8 (1.0)	169.1 (1.1)	----	----	----
	NO_2	----	----	398.7 (0.5)	403.9 (0.4)	407.6 (3.3)
	NO_2+SO_2	----	169.0 (3.1)	399.8 (0.8)	----	407.4 (0.4)
Portland C (fluorocarbon/ phosphate)	SO_2	----	----	----	----	----
	NO_2	----	----	399.3 (0.9)	----	406.5 (0.2)
	NO_2+SO_2	----	----	400.3 (0.4)	----	----
Portland D (fluorocarbon/siloxane)	SO_2	167.5 (0.6)	169.1 (0.7)	----	----	----
	NO_2	----	----	399.7 (1.4)	----	407.7 (0.6)
	NO_2+SO_2	----	168.7 (0.9)	400.5 (0.9)	----	407.6 (0.4)

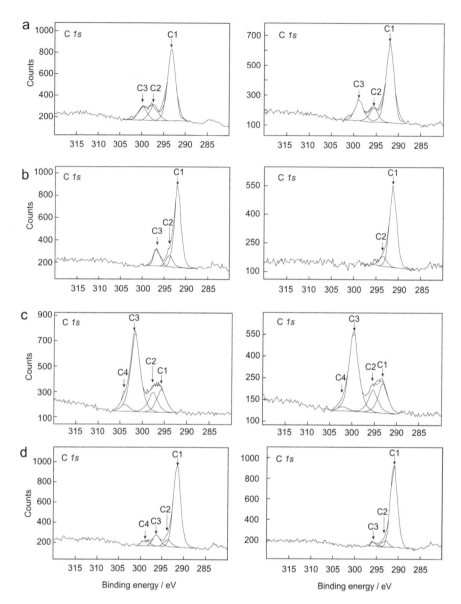

Figure 5.3 Carbon 1s peak fits from Monk's Park samples before and after exposure to NO_2+SO_2; (a) coating A (fluorocarbon); (b) coating B (siloxane/fluorocarbon); (c) coating C (fluorocarbon/ phosphate); (d) coating D (fluorocarbon/siloxane).

falls in concentration of the active constituents of the coatings following exposure to SO_2 or NO_2+SO_2 were seen with coatings A (fluorocarbon), fluorine decreasing and D (fluorocarbon/siloxane), fluorine and silicon decreasing while calcium increased. Interestingly, with the samples coated with B (siloxane/ fluorocarbon), while the surface concentration of silicon decreased, that of fluorine increased. It is clear that these coatings are not acting as passive barriers but can actively participate in a complex suite of reactions, which affects their ability to protect the underlying stone from attack. Figure 5.3 presents the peak-fitted carbon 1s regions for the coated Monk's Park samples before and after exposure to NO_2+SO_2 (these spectra have not been charge corrected). For coatings A (fluorocarbon) and D (fluorocarbon/siloxane), we see a reduction in the intensity of the peak due to calcite (C2 for coating A and C3 for coating D, at ~289.6 eV after charging correction) following exposure, as would be expected as this is attacked and converted into the sulphate. This effect is most marked with coating B (siloxane/ fluorocarbon), where the calcite peak is removed completely. However, little change is observed in the spectra from the coating C (fluorocarbon/phosphate) sample.

DISCUSSION

Results from the controlled exposure experiments on uncoated Monk's Park Bath Stone have been reported previously.[18,19,20] However, further careful consideration of the XPS data has enabled us to elucidate a molecular orbital model for the adsorption of NO onto the surface of calcite, $CaCO_3$.

The chemistry of NO is dominated by the valence electron behaviour, and the $2p$ orbitals dictate the bonding characteristics. Studies of NO adsorbed to surfaces have shown that lower binding energy states, $c.399$ eV, represent molecules tilted relative to the surface normal, by of the order of 59°, and with significant back donation into the π^* NO orbitals whereas higher binding energies, $c.401-402$ eV, are

86 STONE DECAY ITS CAUSES AND CONTROLS

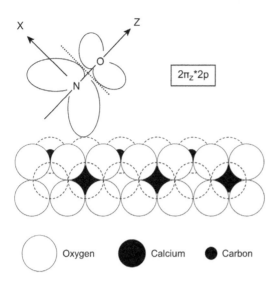

Figure 5.4 The expected limestone surface with adsorbed NO.

indicative of more perpendicular NO with no back donation.[21,22,23] The present measurements showed only one form of adsorbed NO, with a binding energy of $c.400.0(\pm0.3)$ eV.

In the limestone structure, CO_3^{2-} groups and Ca^{2+} ions are held together by electrostatic forces forming an ionic crystal. In the present study, it is assumed that the surface of limestone is occupied by oxygen atoms from the CO_3^{2-} groups. Tilted adsorption of NO onto such a surface is shown schematically in Figure 5.4.

The gas phase photoelectron spectrum of NO was first reported by Siegbahn *et al.* and interpreted using a simple molecular orbital model.[24] After applying appropriate corrections, the average nitrogen 1s binding energy for NO in the gas phase spectrum was found to be 406.5 eV, compared to our value for NO adsorbed on a limestone surface of $c.400.0(\pm0.3)$ eV. This reduction in the measured binding energy may be explained by assuming the transfer of an electron from the substrate to the lower unoccupied electronic levels of adsorbed NO. The nature of the adsorbed species is as depicted in Figure 5.4. It is suggested that the transfer of electron density to the NO molecule via the antibonding $2\pi^*$ orbital leads to an effective weakening of the N-O bond. Since Siegbahn's measurements indicate that the unpaired electron in the $2\pi^*$ orbital resides on the nitrogen atom, this electron transfer is reflected in a lower nitrogen 1s binding energy.

Having shown how the NO reaction product is chemisorbed onto the calcite surface, it is possible to consider the way in which this might affect the progress of the reaction. It may be viewed as a reservoir of NO for oxidation back to NO_2 and, therefore, able to accelerate sulphation of the limestone, even if further NO_2 is not available from the gas phase. If it was not strongly bound to the calcite surface it would be more likely to be reoxidized back to NO_2 in the gas phase away from the stone surface and, therefore, less likely to cause further deterioration of the stone. This helps to explain the deleterious effect that NO_2 has in enhancing the rates of stone corrosion of the lessening concentrations of SO_2 in our urban atmospheres.

Our results have shown that the coatings do not provide a uniform coverage over the stone, but, instead, leave areas uncoated and vulnerable to attack from NO_2 and SO_2. The extent of surface coverage, and, thus, initial protection of the stone was a function of the viscosity and composition of the coating material.

References

1 Allen G.C. and Beavis J. (1996) Building on the Past. *Chemistry in Britain* 32: 24–28.
2 Johansson L-G., Lindqvist O. and Mangio R.E. (1988) Corrosion of Calcareous Stones in Humid Air Containing SO_2 and NO_2. *Durability of Building Materials* 5: 439–449.
3 Johansson L-G. (1990) Synergistic Effects of Air Pollutants on the Atmospheric Corrosion of Metals and Calcareous Stones. *Marine Chemistry* 30: 113–122.
4 Haneef S.J., Jones M.S., Johnson J.B., Thompson G.E. and Wood G.C. (1993) Effects of Air Pollution on Historic Buildings and Monuments (1986–1990). Scientific Basis for Conservation: Laboratory Chamber Studies. *European Cultural Heritage Newsletter on Research* 7: 2–10.
5 Allen G.C., El-Turki A., Hallam K.R., McLaughlin D. and Stacey M. (1999) The Degradation of Limestone and Calcite in NO_2 and SO_2 Atmospheres. *Proceedings of the 7th. Euroseminar on Microscopy Applied to Building Materials*. Delft University of Technology: Delft: 331–337.
6 Allen G.C., El-Turki A., Hallam K.R., McLaughlin D. and Stacey M. (2000) The Role of NO_2 and SO_2 on the Degradation of Limestone. *British Corrosion Journal* 35: 35–38.

7   Allen G.C., El-Turki A., Hallam K.R., McLaughlin D. and Stacey M. (2004 in press) The Role of NO2 and SO2 on the Degradation of Limestone. Proceedings of SWAPNET '99, Stone Weathering in Polluted Urban Environments. University of Wolverhampton: Wolverhampton; 13–14. May
8   Allen *et al, op. cit.* (1999)
9   Allen *et al, op. cit.* (2000)
10 Allen *et al, op. cit.* (2004)
11 Allen *et al, op. cit.* (1999)
12 Allen *et al, op. cit.* (2000)
13 Allen *et al, op. cit.* (2004)
14 Velic D. and Levis R.J. (1998) Collision-Induced Deposition (CID) of NO from Pt{111}: A Comparison Between the CID Binding Energy and the Activation Energy for Thermal Deposition. *Surface Science* 396: 327–339.
15 Johansson *et al, op. cit.* (1988)
16 Spedding D.J. (1969) Sulphur Uptake by Limestone. *Atmospheric Environment* 3: 683.
17 Christie A.B., Lee J., Sutherland I. and Walls J.M. (1983) An XPS Study of Ion-Induced Compositional Changes with Group-II and Group-IV Compounds. *Applied Surface Science* 15: 224–237.
18 Allen *et al, op. cit.* (1999)
19 Allen *et al, op. cit.* (2000)
20 Allen *et al, op. cit.* (2004)
21 Finn P. and Jolly W. (1972) The Nitrogen 1s Binding Energies of Transition Metal Nitrosyls. *Inorganic Chemistry* 11: 893–895.
22 Roberts M.W. and Smart R.StC. (1980) XPS Studies of Donor and Acceptor Chemisorption of NO and CO on Nickel Oxide Surfaces. *Surface Science* 100: 590–604.
23 Lindsay R., Baumgartel P., Terborg R., Schaff O., Bradshaw A.M. and Woodruff D.P. (1999) Molecules on Oxide Surfaces: A Quantitative Structural Determination of NO Adsorbed on NiO(100). *Surface Science* 425: L401–L406.
24 Siegbahn K., Nordling C., Johansson G., Hedman J., Heden P.F., Hamrin K., Gelius U., Bergmark T., Werme L.O., Manne R. and Baer Y. (1969) *ESCA Applied to Free Molecules.* University of Uppsala: Uppsala.

6 Weathering of Sandstone Sculptures on Charles Bridge, Prague: Influence of Previous Restoration

R. PŘIKRYL, J. SVOBODOVÁ
and D. HRADIL

ABSTRACT

Seventy samples of weathering crusts from three sculptures placed on medieval Charles Bridge (Prague, Czech Republic) have been analysed. Microscopic examination of polished cross-sections of crusts in visible and ultraviolet light was combined with electron microscopy and analytical (x-ray diffraction and microanalysis) studies. Petrographic methods (optical microscopy, computer-assisted image analysis) were used to determine source localities and type of the stone from which the individual sculptures were carved.

The stone material of sculptures (fine-grained quartz sandstones cemented with clay matrix) is covered by μm–mm thick grey to black crusts composed mostly of sulphates. Although gypsum is the pervasive mineral, other neo-formed phases like boussingaultite (NH_4-Mg sulphate), NH_4-Zn sulphate, palmierite (K-Pb sulphate), syngenite (K-Ca sulphate), jarosite (K-Fe sulphate), barite or Ba celestine (Ba±Sr sulphates), and calcite (Ca carbonate) dominate in some places.

Chemical composition of original stones used for the sculptures does not ensure sufficient supply of most of the species (K^+, Ca^{2+}, Zn^{2+}, Pb^{2+}, Ba^{2+}, Sr^{2+}, NH_4^+) comprising the minerals. The only possible explanation for their presence is the past use of chemicals for painting, cleaning, paint removal and conservation of the stone. Sulphur and carbon oxides are predominantly derived from the polluted urban atmosphere.

INTRODUCTION

The formation of weathering crusts on monumental stones and change of their colour (surface blackening) is an undesirable process occurring in polluted urban areas[1,2] and is considerably affected by geographic location.[3] Weathering crusts on building stones are often attributed to the deposition of pollutants,[4,5] to the formation of new phases from material mobilized from the stone,[6] and to microbial activity.[7,8] The composition and actual appearance of crusts depend significantly on the exposure of the stone surface.[9] The formation of gypsum crusts on limestones and marbles[10,11] or on clastic sedimentary rocks containing calcite cement[12,13] is a common process and is quite obvious in polluted atmospheres with excess SO_2. The sulphate crusts on rocks containing no carbonates are quite unusual and their formation is not fully understood although mechanisms like the sulphation of atmospheric carbonate particles[14] or evaporation from salts drawn up from soil[15] have been suggested.

This study aims to find the main causes of the formation of weathering crusts on stones with mineralogical and chemical compositions that do not primarily favour formation of sulphate minerals. The crusts were sampled from three original sculptures on the medieval Charles Bridge over the Vltava River in Prague:

♦ St. Wenceslas (1858), (Figure 6.1a)
♦ St. John de Matha, St. Felix de Valois and the Blessed Ivan (1714), (Figure 6.1b)
♦ St. Vincent Ferrer and St. Procopius (1712), (Figure 6.1c).

WEATHERING OF SANDSTONE SCULPTURES ON CHARLES BRIDGE 91

Figure 6.1 Statues on the Charles Bridge, Prague: (a) St. Wenceslas (1858); (b) St. John de Matha, St. Felix de Valois and the Blessed Ivan (1714); (c) St. Vincent Ferrer and St. Procopius (1712).

Detailed analytical study of those crusts and substrate rocks provided possible explanation of the origin of sulphate-rich crusts on quartz sandstones.

STONE MATERIAL

Macroscopic inspection of the sculptures shows them to have been carved from fine-grained quartz sandstones, placed on bases carved from different types of sandstones and arkoses.

The stone of the sculptures and their bases was carefully sampled in order to obtain material for detailed microscopic study. Samples were fixed in epoxy resin with fluorescent dye to increase contrast between clasts and pore space. Conventional thin sections were then prepared. Thin sections were inspected using an optical microscope for qualitative description. Optical microscopy confirmed that all stones used for the sculptures are fine-grained Cenomanian (Upper Cretaceous) quartz sandstones with clay matrix (Table 6.1). None of the samples contained carbonates either in the matrix or as clasts. The surface of the stone is covered by grey to black compact crust (Figure 6.2).

Table 6.1 Mineralogical composition and source localities of studied sculptures.

	St. Vincent	St. John	St. Wenceslas
Dominant clastic minerals	Quartz	Quartz	Quartz
Minor clastic minerals	K-feldspar, muscovite	K-feldspar, plagioclase, muscovite	K-feldspar, muscovite
Matrix	Kaolinite, goethite, organic matter	Kaolinite	Kaolinite
Average grain size [mm]	0.08	0.08	0.08
Source locality	Surroundings of Mšené-lázně	Surroundings of Mšené-lázně	Surroundings of Mšené-lázně

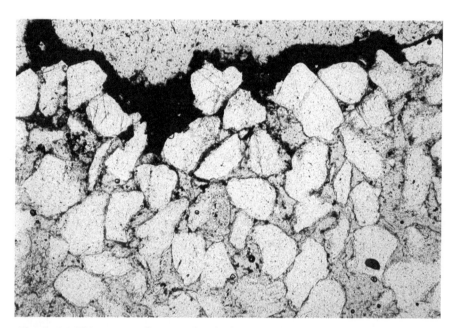

Figure 6.2 Thin-section showing detail of the relatively compact structure of the sandstone used in the sculptures (Picture width 1.35 mm).

quantitative data that allowed discrimination of the source localities of the stone material (Table 6.1). The procedure involves outlining and digitizing mineral/clast boundaries, and image analysis (SIGMASCAN, Jandel Scientific, USA). The results (grain size distribution, grain shape) were compared to the data obtained from similar analyses of rocks taken from possible source localities in the broad surroundings of Mšené-lázně.The bases of studied sculptures were carved from 'Hořice' quartz sandstone (another type of fine to medium-grained Upper Cretaceous sandstone) or from Carboniferous arkoses. Neither of these stones contain calcite.

CRUSTS

The surfaces of sculptures do not exhibit any signs of material loss such as flaking, spalling, granular disintegration or alveolar weathering (Figure 6.3). Weathering is predominantly manifested as blackening and surface crusting. Crusts do not cover entire surfaces of the sculptures, but concentrate on certain places such as in rain shadow areas. Biological attack by algae and lichens is limited in extent. The visible weathering of the sculptures' bases is flaking and spalling of the stone surface.

Samples of the crusts (18 from St. Wenceslas, 38 from St. John, 14 from St. Vincent) were collected from different places in order to obtain representative sets of samples to characterize different depositional environments (sheltered areas, exposed surfaces etc.). The samples were mounted in polyester resin and polished perpendicular to the crust surface. These cross-sections were examined and photographed in reflected visible and ultraviolet light by optical microscope.

This method provided information on stratigraphy and composition of the crusts. Most of the crusts show a typical vertical profile characterized by a surface layer of translucent (probably pure) sulphate crystals and a dark bottom layer (Figure 6.4a). The arrangement of crystals is botyroidal (pseudoradial) and more or less

94 STONE DECAY ITS CAUSES AND CONTROLS

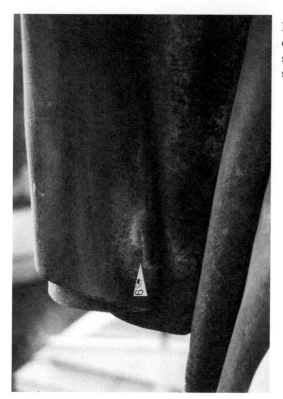

Figure 6.3 Detail of the extent and development of surface crust in a relatively sheltered location.

perpendicular to the surface. The surface layer is occasionally composed of thin multilayers of sulphates as observed on the St. Vincent sculpture (Figure 6.4b).

The upper layer of pure sulphates is underlined by a more complex mixture of minerals derived from the underlying original stone (quartz, clay minerals), deposited solid dust particles and newly formed sulphates. This layer can be distinguished from the upper one by its dark colour. Moreover, very thin layers of colour were found in some samples indicating previous restoration. The microscopic study also focused on the presence of original paint on the stone surfaces. The remnant of several pigmented layers was observed only in one sample taken from the St. John sculpture.

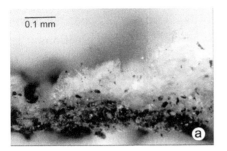 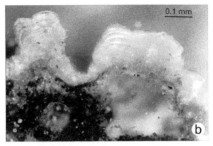

Figure 6.4 Cross-sections showing the microstratigraphy on crust samples from the sculpture of St. Vincent; (a) upper layer of translucent gypsum with a lower dark layer comprising sulphates, quartz, dust particles etc.; (b) surface sulphate multilayers.

X-RAY DIFFRACTION

X-ray powder diffraction (XRD) was employed to obtain detailed data on mineralogical composition of studied crusts. Part of each crust was homogenized by powdering in an agate cup. Samples with macroscopically visible quartz were sieved. The coarse fraction was further separated into dark and light fractions. Each fraction was analysed separately.

Measurements were conducted using a SIEMENS D-5005 with CuKa radiation, secondary monochromator, voltage 40 kV, current 30 mA, degree range 2Theta 3–90°, step 0.02° per 2.4 seconds. The raw data were processed by a ZDS for Windows program[16] and a diffraction pattern database.[17]

The mineralogical composition of the sampled crusts was found to be dominated by sulphates as weathering minerals and by quartz, kaolinite and feldspars as minerals derived from the underlying sandstone (Table 6.2). Additionally, the weathering crusts of each sculpture exhibit specific mineralogies. The sculpture of St. Wenceslas shows a dominant presence of gypsum in all samples and rare occurrences of boussingaultite and/or mohrite (in two samples only). These double sulphate hexahydrates of an ammonium ion and one divalent cation show conforming structural features causing similarities in their diffraction records. Because of possible non-stoichiometry that also can affect the diffraction peak positions it is difficult to specify the cation prevailing in the structure. Owing to the fact that the presence of magnesium has been undoubtedly confirmed by electron microanalysis (EDAX, see below), the mineral phase of boussingaultite is considered to be dominant. The presence of mohrite, in which magnesium is substituted by divalent iron, remains uncertain.

The greatest mineralogical diversity was found in weathering crusts of the group of sculptures of St. John de Matha. Gypsum dominates in most of the samples but other sulphates were found in important quantities. Beside boussingaultite and speculative mohrite, large crusts of isostructural zinc ammonium sulphate hexahydrate were found in one sample (Figure 6.5) and the presence of Zn was confirmed by EDAX (see below). Double sulphates of potassium and lead (palmierite) and potassium and calcium (syngenite) were also recorded. The alunite group of minerals is represented by jarosite found in one sample in a significant amount. Although there is no clear evidence for alunite itself, the association of elements K-Al-S was often identified by EDAX. Moreover, barite, small amounts of carbonates like calcite, aragonite and speculative barytocalcite and nyererite were traced. In one sample, a small amount of white lead (Pb carbonate hydroxide) was measured (Figure 6.6).

A relatively broad mineralogical diversity was also observed in the crusts from the St. Vincent group of sculptures. In addition to common gypsum, palmierite and barium celestine were recorded in three samples.

Some phosphates occur in crusts, especially from the St. John and St. Vincent sculptures, but their precise mineralogy was not determined due to their poor crystallinity causing the broadening of diffraction peaks. This is a common obstacle and other methods like FTIR might provide a more definitive identification.

WEATHERING OF SANDSTONE SCULPTURES ON CHARLES BRIDGE 97

Table 6.2 Summary of mineralogical composition of weathering crusts (perceptual occurrence of each well-identified phase is noted in parentheses, speculative/uncertain phases are in italics)

	St. Vincent (14 samples)	St. John (38 samples)	St. Wenceslas (18 samples)
		Neo-formed phases	
SULPHATES			
– hydrated $A^{II}SO_4.xH_2O$	Gypsum (100%) (A = Ca; x = 2)	Gypsum (92%) (A = Ca; x = 2)	Gypsum (100%) (A = Ca; x = 2)
$A^I_2B^{II}(SO_4)_2.xH_2O$ (including picromerite group hexahydrates)		Boussingaultite (10%) (A = NH$_4$; B = Mg; x = 6) Ammonium zinc Sulphate (3%) (A = NH$_4$; B = Zn; x = 6) *Mohrite* (A = NH$_4$; B = Fe^{II}; x = 6) Syngenite (3%) (A = K; B = Ca; x = 1)	Boussingaultite (11%) (A = NH$_4$; B = Mg; x = 6) *Mohrite* (A = NH$_4$; B = Fe^{II}; x = 6)
– anhydrous $A^{II}SO_4$ (barite group) $A^I_2B^{II}(SO_4)_2$	Ba celestine (21%) (A = Ba, Sr) Palmierite (21%) (A = K, Na; B = Pb)	Barite (3%) (A = Ba) Palmierite (6%) (A = K, Na; B = Pb)	
– anhydrous with hydroxyl group $A^IB^{III}_3(SO_4)_2(OH)_6$ (alunite group)		Jarosite (3%) (A = K; B = Fe^{III}) *Alunite* (A = K; B = Al)	
CARBONATES – anhydrous		Calcite (13%), Aragonite (3%) CaCO$_3$ *Barytocalcite* BaCaCO$_3$ *Nyererite* Na$_2$Ca(CO$_3$)$_2$	
PHOSPHATES	*unspecified*	*unspecified*	
PAINT REMNANTS		Lead white 2PbCO$_3$.Pb(OH)$_2$	
DUST/FLY ASH PHASES	Glass phase (Si, Al, Fe, Ti, K) rutile (TiO$_2$)	Glass phase (Si, Al, Fe, Ti, K) Fe oxide (e.g. magnetite)	Glass phase (Si, Al, Fe, Ti, K)
STONE ORIGINATED PHASES	Quartz, kaolinite	Quartz, kaolinite, K-feldspar, plagioclase, mica	Quartz, kaolinite, goethite, plagioclase

98 STONE DECAY ITS CAUSES AND CONTROLS

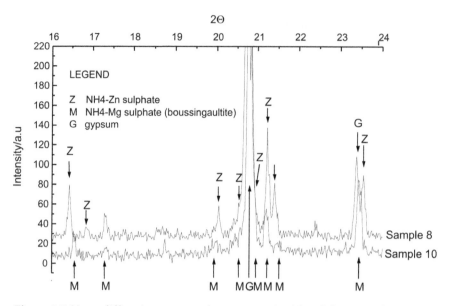

Figure 6.5 X-ray diffraction patterns from ammonia-rich sulphates on the St. John sculpture, based on substitution of inorganic cations in the crystal structure (Zn – sample 8; Mg – sample 10).

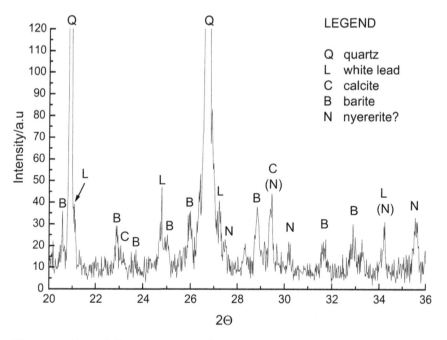

Figure 6.6 X-ray diffraction pattern of crust sample from the sculpture of St. John showing the presence of carbonates, barite and relicts of white lead.

SCANNING ELECTRON MICROSCOPY

Detailed examination of crust surfaces was conducted by scanning electron microscopy (SEM) combined with energy dispersive x-ray analysis (EDAX). The crusts were covered with gold (for SEM only) or with graphite (for SEM/EDAX). The samples were analysed and photographed on a JEOL JXA 50A+EDAX 9400 apparatus and photographed using a Philips XL30.

The morphology of the crusts shows significant qualitative differences between the sculptures. Crusts on St. Vincent exhibit the best-developed crystallographic forms of gypsum with few grains of dust particles (Figure 6.7). The surface of gypsum crystals is more obscured by dust particles and amorphous layers (Figure 6.8) in the crusts from St. Wenceslas. The most uneven crust surfaces were observed on St. John. This is probably due to the presence of other sulphates such as ammonium zinc sulphate (Figure 6.9) or boussingaultite that tend to develop solid and polygonally cracked surfaces rather than ordinary crystal shapes like gypsum.

Dust particles from the crust surface (Figure 6.10) were analysed by EDAX. They typically comprise a silicate matrix including Al, Fe, Ti, and K. Contents of Fe and Ti are variable, in some cases pure oxides have been analysed (magnetite was recorded by XRD). This chemistry corresponds to the composition of coal-derived fly-ash particles reported in other studies.[18] These particles cannot be fully detected by XRD due to the dominance of the matrix.

SOURCES OF ELEMENTS IN THE CRUSTS

The stone material of the three groups of sculptures is quartz sandstone with clay matrix. The potential source locality of these sandstones was determined as the surroundings of Mšené-lázně village about 30 km NW from Prague. XRD study of original stone from the

Figure 6.7 Scanning electron micrograph of crust sample from the sculpture of St. Vincent showing extensive gypsum coverage.

Figure 6.8 Scanning electron micrograph of crust sample from the sculpture of St. Vincent showing a surface covered by dust particles and other amorphous material.

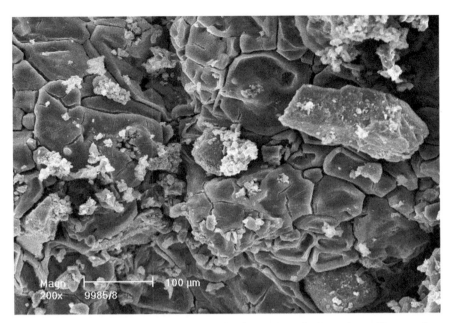

Figure 6.9 Scanning electron micrograph of crust sample from the sculpture of St. John showing a desiccated surface morphology rich in ammonium zinc sulphate.

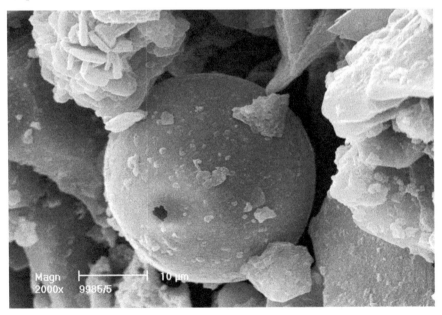

Figure 6.10 Scanning electron micrograph of crust sample from the sculpture of St. John showing detail of surface fly-ash deposit composed of a silicate matrix with Al, Fe, Ti and K as identified by EDX analysis.

quarry revealed the presence of quartz, K-feldspar, traces of Ca-plagioclase and clay matrix minerals (kaolinite and small amount of illite). The original stone (Table 6.3) shows little excess calcium that occurs probably partly in plagioclase, and partly adsorbed on the clay surfaces. Sulphur and divalent iron can hypothetically form pyrite in trace amounts. Other elements – Mg, Zn, Pb, Ba – or compounds – NH_4^+– that accompany the formation of weathering crusts are not present. Amounts of Ca, K, SO_2 and CO_2 measured are not sufficient to form the crusts. It has been suggested that the primary source of elements within crusts on Charles Bridge main structure is rock-forming minerals,[19] this source has been excluded for the sculptures due to the extreme variability of crust mineralogy and the strict homogeneity of stone substrate. The extent of mineral weathering (especially of feldspars) is observable to the same extent in stone from sculptures and analogous rocks from natural outcrops.

The only possible explanation is that crustal elements and compounds were introduced to the stone later, either as contaminants from the polluted atmosphere (SO_2, CO_2) or during conservation of those sculptures or from the biosphere.

Four possible sources of elements are suggested. The first group (Pb, Zn) is probably related to the presence of original polychromy on the stone surface or to chemical cleaning. The second group of elements (K) might be related to chemicals used for removal of paint. The third group of elements (Ca, Mg, Ba, ±Zn, ±NH_4^+) was possibly introduced by chemical cleaning and consolidation of the stone. Finally, some elements (P) or compounds (NH_4^+) may be related to bird droppings.

The presence of polychromy on two older sculptures is documented by archive materials as well as by occasional observations of paint fragments in surface layers. The formation of rare ammonium zinc sulphate [$(NH_4)_2Zn(SO_4)_2.6H_2O$] (St. John) and palmierite [$K_2Pb(SO_4)_2$] (St. John and St. Vincent) also gives evidence of paint remnants (Zn originated from zinc white, Pb originated from red lead or white lead; relicts of which were also detected by XRD) that reacted with other compounds and atmospheric SO_2. Another

Table 6.3 Chemical composition of rocks analogous to the stone material of studied sculptures. The rock was analysed by XRF (Gematest Ltd., analyst Ing. A. Manda, 1999). The rocks were sampled in abandoned quarries in Mšené-lázně area – surroundings of Mšené-lázně (1), Budyně (2) and Přestavlky (3) where fine-grained quartz sandstone has been quarried for several centuries.

	Mšené-lázně (%)	Budyně (%)	Přestavlky (%)
SiO_2	98.04	98.21	98.26
TiO_2	0.10	0.07	0.06
Al_2O_3	0.61	0.52	0.62
Fe_2O_3	0.13	0.28	0.08
FeO	0.11	0.22	0.10
CaO	0.12	0.14	0.14
MgO	<0.01	<0.01	<0.01
MnO	<0.01	<0.01	<0.01
Na_2O	<0.01	0.02	<0.01
K_2O	0.20	0.32	0.25
P_2O_5	0.09	0.05	0.05
SO_3 total	0.07	0.09	0.01
CO_2	<0.10	<0.10	<0.10
Loss on combustion	0.42	0.30	0.25
SO_3 sulphur	0.05	0.01	<0.01
Cl^-	0.01	0.02	0.01
F^-	0.04	0.05	0.06

possible source of Pb and Zn is from Pb fluates and from Zn fluates, respectively.

Both older sculptures (St. John, St. Vincent) also show frequent sulphates with potassium (syngenite [$K_2Ca(SO_4)_2 \cdot H_2O$], jarosite [$KFe_3(SO_4)_2(OH)_6$] and palmierite [$K_2Pb(SO_4)_2$]). Although small amounts of potassium could be mobilized from silicates in the stone, most of the potassium was probably introduced to the stone through potassium hydroxide [KOH]. This chemical is a common alkaline cleaner employed by restorers to saponify the polychromy. Another possibility is impregnation of the stone by potassium water glass.

The presence of calcium (forming pervasive gypsum crusts) can be explained by the use of limewater for the treatment (consolidation) of the stone surface. Although this technique is practised mostly for limestones,[20] it has been applied to other porous stones, such as sandstones. The limewater [$Ca(OH)_2$] easily reacts with atmospheric SO_2 resulting in gypsum [$CaSO_4.2H_2O$] formation. The limewater technique was probably applied on all studied sculptures as gypsum dominates in all crusts.

Boussingaultite [$(NH_4)_2Mg(SO_4)_2.6H_2O$] is another sulphate found in crusts on the St. Wenceslas and St. John sculptures. NH_4^+ has been probably introduced by using ammonium hydroxide ('ammonium water') [$NH_4(OH)$] as a cleaning agent or ammonium hydrogen carbonate [$NH_4(HCO_3)$] employed for the softening of sulphate crusts. Another possible source of NH_4^+ is urea [NH_2CONH_2] that is used as a catalyst in the 'barium water' technique (see below). The third possible (and most speculative) explanation is that they originate from bird droppings. Bird droppings were observed on all three sculptures while boussingaultite-mohrite was found only on two of them.

Magnesium can be introduced to the stone by application of fluates that are used as consolidant agents in carbonate rocks[21] and inhibitors of salt action in any stones. The use of fluates has been documented since the end of nineteenth century and they belong to the common inorganic chemicals used in restoration practice.

Barium minerals (barite [$BaSO_4$], and Ba celestine [$BaSr(SO_4)_2$]) were also recorded in the crusts of the two older sculptures. They can be related to the consolidation of the stone using a 'barium water' technique (barium hydroxide) [$Ba(OH)_2$]. This should react with CO_2, and urea as a catalyst, to form less soluble $BaCO_3$.[22] Our study shows that this reaction is modified in polluted atmospheres with excess SO_2 resulting in preferential sulphate formation.

The SEM/EDAX study revealed the presence of an amorphous dense crust composed of silica on the surface of the stone from sculptures' bases. This implies some use of waterglass for base consolidation that results in the formation of amorphous silica [$SiO_2 \cdot nH_2O$] layer. The resulting effect of this layer is the spalling and flaking of the stone surface that is quite common on the bases.

CRUST ORIGIN

The explanation of crust formation on the sampled sculptures on Charles Bridge can also be supported by the time period of their origin. The sculptures of St. John and St. Vincent were carved at the beginning of the eighteenth century. They were certainly painted and thus partly protected from weathering processes. During the nineteenth century, a change in the care of cultural heritage allowed trials to remove all 'artificial' layers including polychromy. The polychromy has been extensively removed by chemical cleaning agents[23] and stone of the sculptures exposed. The chemicals used introduced uncommon elements to the stone structure. Some elements liberated during polychromy removal were also captured in the pore system of the stone.

After a certain time period, blackening and crust formation occurred. These undesirable effects were treated by other chemicals in order to obtain a clean unweathered surface. The exposed sculptures were then chemically impregnated to protect them[24] and more elements and compounds were introduced to the stone structure. The result of this treatment is the complex mineralogy of the weathering crusts (dominant gypsum accompanied with palmierite, jarosite, syngenite, bousingaultite, zinc ammonium sulphate and others).

The youngest sculpture in the sample set, St. Wenceslas, suggests another history. This sculpture originated in the second half of the nineteenth century, i.e. after the period in which the painting of stone sculptures was practised. This sculpture was subjected only to the cleaning of black patinas and weathered crusts, followed by consolidation. This could explain the relatively simple mineralogical composition of the crusts (dominant gypsum, rare boussingaultite) on the St. Wenceslas sculpture.

CONCLUSIONS

Detailed analytical study of the mineralogy of sulphate crusts observed on quartz sandstone sculptures demonstrates their anomalous origins. As the mineralogical composition of the crusts cannot be related to the chemical or mineralogical composition of the original stone of the sculptures they must have originated in part from other sources. The broad variety of crust mineralogy on older sculptures is caused by a complex interaction between the remnants of polychromy, chemical agents used for its removal and chemical agents used for later cleaning and surface consolidation. The mineralogical composition of crusts on the younger sculpture is, however, more simple than the others. The compounds active in the formation of the crust came only from chemicals used for stone cleaning and conservation. The sulphur, necessary for sulphate formation, comes in all cases from atmospheric SO_2.

The results of this study show that certain chemicals used for the restoration of stone sculptures can provoke later formation of weathering crusts with anomalous mineralogies. However, the lack of precise written restoration reports (especially of notes on applied chemicals) means that explanations of crust mineralogy must remain speculative.

Acknowledgements

The analytical work was financially supported through a research grant from the Czech Ministry of Education to Assoc. Prof. Petr Siegl (grant No. MSM 520000001). The authors are very grateful to Assoc. Prof. Petr Siegl, Ondřej Šimek and Jakub Gajda (School of Art Restoration) for help with sampling. The XRD in the Institute of Inorganic

Chemistry, Academy of Sciences of the Czech Republic, was operated by Petr Bezdička and Antonín Petřina. The scanning electron microscope, Geological Institute of the Czech Academy of Sciences, was operated by Anna Langrová and Zuzana Korbelová. Thanks must also go to Jan Šubrt and Sněžana Bakardžieva, Institute of Inorganic Chemistry of the Czech Academy of Sciences. The Gallery of the City of Prague is acknowledged for permitting access to the sculptures. The final form of this study benefited from comments of two anonymous reviewers.

References

1. Winkler, E.M. (1994) *Stone in Architecture. Properties, Durability.* Springer-Verlag, Berlin.
2. Vendrell-Saz, M., Krumbein, W.E., Urzi, C. and Garcia-Valles, M. (1996) Are patinas of Mediterranean monuments really related to the rock substrate? *Proceedings of the 8th International Congress on deterioration and conservation of stone*, 30 September–4 October 1996, Berlin, Germany 2.
3. Halsey, D.P., Dews, S.J. Mitchell, D.J. and Harris, F.C. (1996) The black soiling of sandstone buildings in the West Midlands, England: regional variations and decay mechanisms. In Smith, B.J. and Warke, P.A. (eds) *Processes of Urban Stone Decay*. Donhead Publishing Ltd, London; 53–65.
4. Sabbioni, C. and Zappia, G. (1992) Decay of sandstone in urban areas correlated with atmospheric aerosol. *Water, Air and Soil Pollution* 63:
5. Nord, A.G., Svardh, A. and Tonner, K. (1994) Air pollution levels reflected in deposits on building stones. *Atmospheric Environment* 28: 2615–2622.
6. Nord, A.G. and Ericson, T. (1993) Chemical analysis of thin black layers on building stone. *Studies in Conservation* 38: 25–35.
7. Wakefield, R.D., Jones, M.S. and Forsyth, G. (1996) Decay of sandstone colonised by an epilithic algal community. In Smith, B.J. and Warke, P.A. (eds) *Processes of Urban Stone Decay*. Donhead Publishing, London; 88–97.
8. Machill, S., Althaus, K., Krumbein, W.E. and Steger, W.E. (1997) Identification of organic compounds extracted from black weathered surfaces of Saxonean sandstones, correlation with atmospheric input and rock inhabiting microflora. *Organic Geochemistry* 27(1–2): 79–97.
9. Steger, W.A. and Mehner, H. (1998) The iron black weathering crusts on Saxonian sandstones investigated by Mössbauer spectroscopy. *Studies in Conservation* 43: 49–58.
10. Jaynes, S.M. and Cooke, R.U. (1987) Stone weathering in southeast England. *Atmospheric Environment* 21(7): 1601–1622.

11 Winkler, E.M. (1988) Weathering of crystalline marble. In P.G. Marions and G.C. Koukis (eds) *Engineering Geology of Ancient Works, Monuments and Historical Sites*. Balkema, Rotterdam; 717–721.
12 Pavía Santamaría, S., Cooper, T.P. and Caro Calatayud, S. (1996) Characterisation and decay of monumental sandstone in la Rioja, Northern Spain. In Smith, B.J. and Warke, P.A. (eds) *Processes of Urban Stone Decay*. Donhead Publishing, London; 125–132.
13 Garcia-Valles, M., Blázques, F., Molera, J. and Vendrell-Saz, M. (1996) Studies of patinas and decay mechanisms leading to the restoration of Santa Maria de Montblanc (Catalonia, Spain). *Studies in Conservation* 41.
14 Garcia-Valles, M., Molera, J. and Vendrell-Saz, M. (1997) A siliceous sandstone in an urban environment: the decay and cleaning of the Church Betlem (Barcelona, Catalonia). *International Journal for Restoration of Buildings and Monuments* 3(5): 469–485.
15 Williams, R.B.G. and Robinson, D.A. (2000) Effects of aspect weathering: Anomalous behaviour of sandstone gravestones in southeast England. *Earth Surface Processes and Landforms* 25: 135–144.
16 Ondruš, P. (1997) *ZDS – software for X-ray powder diffraction analysis*. ZDS Systems Inc., Prague.
17 JCPDS (1999) *Powder Diffraction File*, PDF-2, International Centre for Diffraction Data, Newtown, Pennsylvania, USA.
18 Tazaki, K., Fyfe, W.S., Sahu, K.C. and Powell, M. (1989) Observations on the nature of fly ash particles. *Fuel* 68: 727–734.
19 Pospíšil, P., Locker, J., Gregerová, M. and Sulovský, P. (1994) Engineering-geological aspects of reconstruction of Charles Bridge in Prague. *Proceedings of the 7th International IAEG Congress*, 5–9 September 1994, Lisbon, Portugal, Volume 5, Theme 4: 3627–3634.
20 Winkler, *op. cit.*(1994)
21 Kessler, L. (1883) Sur un procédé du durcissement des pierres calcaires tendre au moyen des fluosilicates a base d'oxydes insolubles. *C.R. Acad. Sci. Inst. Fr.* 96: 1317–1319.
22 Winkler, *op. cit.*(1994)
23 Nejedlý, V. (1994) Notizen zu historischen Oberflächenbehandlungen von Steinbildwerken in böhmischen Kronländern. *Zeitschrift für Kunsttechnologie und Konservierung* 8(2): 255–278.
24 Ibid.

7 Experimental Weathering of Rhyolite Tuff Building Stones and the Effect of an Organic Polymer Conserving Agent

Á. TÖRÖK, M. GÁLOS and
K. KOCSÁNYI-KOPECSKÓ

ABSTRACT

Miocene rhyolite tuff has been used as a building material and ornamental stone in north-eastern Hungary for several hundreds of years. Rhyolite tuff buildings including the Castle of Sárospatak now show severe signs of deterioration. In this study we present the results of petrophysical and mineralogical tests performed to determine the durability of rhyolite tuff and assess performance of a conserving agent, a polymethyl methacrylate (PMMA). For experimental weathering disc-shaped samples (5 cm in diameter) were used and tested after freeze-thaw cycles and magnesium sulphate crystallization cycles. Resistance to weathering and the effect of the PMMA consolidant were measured by detecting changes in physical properties including surface roughness, tensile strength, Duroscope rebound, water absorption and ultrasound velocity. Mineralogical analyses showed that in the non-crystalline glassy groundmass of the pumice-rich tuff quartz, biotite and sanidine are the most frequent minerals. In naturally weathered tuff samples clay minerals (montmorillonite, illite, chlorite) and calcite were detected while it was not possible to measure similar changes in experimentally weathered samples by XRD. Rhyolite tuff with open voids of more than 20% volume has a high water absorption capacity and strength parameters show a drastic drop after water saturation, i.e. compressive strength becomes less than half. Due to salt weathering or freeze-thaw cycles a spongy tuff surface is developed. The PMMA consolidant strengthens the tuff by forming polymers within the voids. In restoration works saturation of the outer stone surface is the common practice. By applying this method of consolidation PMMA forms a thin load-bearing but permeable crust on the tuff surface. The binding force of the crust is insufficient since additional freeze-thaw cycles or magnesium salt crystallization disrupts this consolidated crust. Consequently the application of PMMA consolidant on exposed stone surfaces of water sensitive rocks, such as this rhyolite tuff is not recommended. Sandstones and limestones are more often treated by organic polymers, but special attention is needed to detect colour changes and the bond between the consolidated crust and host rock to avoid possible adverse effects of such treatment.

INTRODUCTION

Recent studies of stone weathering using experimental slabs have concentrated mostly on salt and frost weathering of sedimentary rocks such as limestones and sandstones.[1,2,3,4,5] Much less attention has been paid to igneous rocks and especially tuffs. This paper intends to bridge this gap by illustrating the results of experimental weathering tests on rhyolite tuff slabs. The degree of weathering is assessed by physical parameters and mineralogical changes.

The other topic that is discussed in this paper is whether a conserving agent can prevent or slow down the weathering of the rhyolite tuff. To investigate this question a polymethyl methacrylate (PMMA, known as Paraloid B72) conserving agent, a consolidant, was tested. The PMMA was chosen for the following reasons:

- during previous restoration works some tuff surfaces were treated by this agent in North Hungary and their effect has not yet been analysed;
- there is a proposal to use PMMA for conserving rhyolite tuff ornaments in current restoration works
- methyl methacrylates are relatively new conserving agents that are mostly applied to limestones and sandstones[6] and there are no previous studies dealing with the effect of this agent on tuffs.

The effectiveness of PMMA conserving agent was assessed by measuring several petrophysical properties of untreated and treated tuff test samples. The conserving agent was applied to fresh and

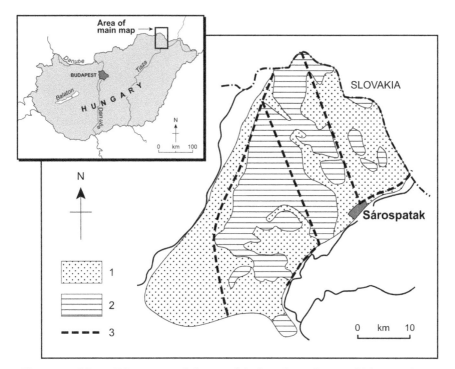

Figure 7.1 Map of Hungary and the simplified geological map of Sárospatak area with the surface occurrence of rhyolite tuff. (1) rhyolite tuff (Miocene); (2) other volcanic rocks (rhyolite, andesite, dacite, andesite tuff, dacite tuff); (3) major faults (geological map modified after Gyarmati 1977).

experimentally weathered stone surfaces. The latter were produced in laboratory conditions by freeze-thaw cycles or magnesium-sulphate crystallization.

Miocene rhyolite tuff from north-east Hungary was used for the tests. This yellowish porous tuff is easy to cut and carve. Being widespread in the Tokaj Mountains it is the most important building and ornamental stone of Sárospatak region (Figure 7.1) and it was used in Gothic, Renaissance, Baroque and Classical monuments and public buildings. The castle of the historic city of Sárospatak was also constructed from rhyolite tuff. During previous restoration works in the 1980s some parts of castle wall and ornaments were treated by PMMA. No previous studies have been carried out to assess the effectiveness of this treatment.

Thus experimental weathering tests are aimed to show the rate of weathering of untreated rhyolite tuff and its comparison to treated surfaces and the effect of PMMA on the durability of stone surfaces.

METHODOLOGY

The analytical methods used can be divided into mineralogical and petrological analyses and petrophysical tests.

Mineralogical and petrological analyses were carried out to describe the composition, texture and porosity of rhyolite tuff. Special emphasis was put on the difference in the composition of unaltered and weathered specimens. In the first phase samples were described and thin sections were prepared and analysed using a polarizing microscope. Mineralogical composition was also detected by XRD. Powdered samples of 50–100 mg were used in a PHILIPS PW 3710 diffractometer with test conditions of: Cu-anode, generator voltage 40kV, generator current 20mA, angle 2Θ: 5–70°. No sample pre-treatment and no internal standards were used.

For petrophysical tests three large blocks (20–40 kg) of rhyolite tuff were chosen. Cylindrical samples of 5 cm in diameter were cored from the blocks. Conventional petrophysical parameters such as density, compressive and tensile strength, ultrasound velocity, water content and water absorption were determined on cylinders (5 cm in height) in the following petrophysical states: air-dry, water saturated, dried at 105°C, 15 and 25 freeze-thaw cycles (five samples in each state). These tests provided standardized values of durability according to construction industry standards.

Resistance to weathering was also evaluated by using disc-shaped specimens (5 cm in diameter and 2 cm in height). The change in water absorption, tensile strength, ultrasound velocity and density after freeze-thaw cycles (15 cycles) or magnesium sulphate crystallization (2 cycles) were also measured.

The effect of the conserving agent on the rhyolite tuff was analysed on similar disc-shaped specimens. Surfaces, which had been cut, damaged by frost attack (15 freeze-thaw cycles) and damaged by magnesium sulphate crystallization (2 cycles), were used for assessing the mechanism and performance of the conserving agent. To do this, 'weathered' tuff surfaces were generated under laboratory conditions to model situations where the conserving agent is applied to weathered monumental stone. The surface characteristics of cut surfaces, frost- and magnesium sulphate-treated surfaces are given in micro-roughness and were measured to a precision of 0.001 mm.

The effect of a conserving agent can be best explained by the changes in petrophysical properties of the tuff. The changes in mass distribution, density and water absorption provide information on the internal structure of the treated tuff while strength tests describe the load-bearing capacity of treated surfaces. The Duroscope rebound test, a non-destructive method, provides information on mechanical properties of rocks giving a rebound value which correlates with the strength properties of the rock surface.

The tested conserving agent is a transparent organic polymer (PMMA). At first, Paraloid 72 granulate was dissolved in nitrodilutent and after dissolution it was used in a liquid form for surface treatment. The consolidating effect of the agent is related to a polymerization process, a vinyl monomer. The methyl methacrylate (MMA), polymerizes and forms a solid and transparent mass within the pores and on the surface of the rock. The evaporation of the solvent and solidification is a rapid process and takes place within hours. Three different treatment methods were evaluated to assess the penetration depth and the performance of PMMA in rhyolite tuff conservation. The entire sample was put into the conserving agent for 5 minutes resulting in a thin (2–4 mm) but continuous treated

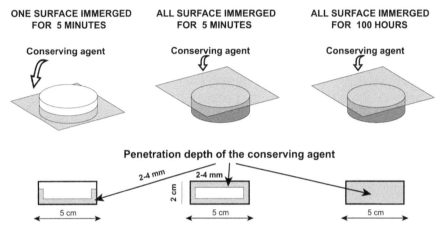

Figure 7.2 Different treatment methods of PMMA conserving agent used for the tests.

Table 7.1 Number of disc-shaped (5 cm in diameter and 2 cm in height) test samples used for assessing the effect of PMMA conserving agent. Reference samples are compared to the samples that were treated in three different ways.

Test conditions	Reference samples	One surface immersed for 5 minutes	All surfaces immersed for 5 minutes	All surfaces immersed for 100 hrs.
Air-dry	4	4	4	4
Water-saturated	4	*	*	*
15 freeze-thaw cycles	4	4	4	4
2 MgSO$_4$ cycles	4	4	4	4
15+5 freeze-thaw cycles	4	4	4	4
2+2 MgSO$_4$ cycles	4	4	4	4

crust. The second series of samples were placed into the agent for 5 minutes in such a way that only the lower part of the disc reached into the agent, i.e. partially submerged. The third series of discs were saturated by the conserving agent (100 hours) leading to the full saturation of the open void spaces (Figure 7.2). For each test 4 samples were used to obtain average and standard deviation values. The results were compared to a reference sample set (Table 7.1).

MINERALOGY, PETROLOGY AND PHYSICAL PROPERTIES OF RHYOLITE TUFF

Mineralogy and petrology

Light creamy to white rhyolite tuff is characterized by significant amounts of groundmass and pumice (56% on average). White porous pumice clasts are major constituents reaching 4–10 mm in diameter. Small mm-sized rock fragments, such as rhyolite or andesite are also present. The most common minerals are grey transparent quartz (12% volume) and sanidine (4% volume). Brownish-black biotite (3% volume) crystals (a few millimetres in size) also form phenocrysts. A minor proportion of plagioclase (albite 1% volume) was also detected. Under the microscope the tuff is characterized by vitroporphyritic texture (Figure 7.3). Minerals, rock fragments and pumice clasts can be regularly arranged forming flow banding or swirling layers of glassy groundmass and aligned phenocrysts. The void space (20% volume) is mostly created by vesicles of pumice (Figure 7.3).

Physical properties

The water content of the tuff is low (1.6% volume), while its water absorption is significant (19.6% volume) indicating a well-connected

116 STONE DECAY ITS CAUSES AND CONTROLS

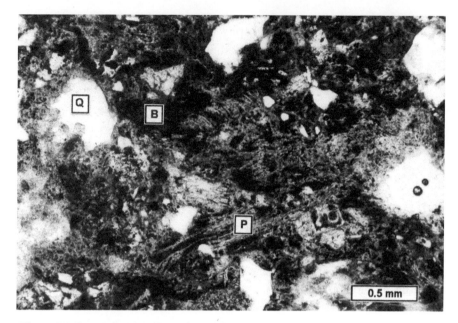

Figure 7.3 Large pumice (P) with vesicles, quartz (Q) and altered limonitic biotite (B) in vitroporphyritic rhyolite (thin-section photograph).

Table 7.2 Petrophysical properties of Sárospatak rhyolite tuff measured on standard cylindrical samples (average values).

Properties/Test conditions	Air-dry	Water saturated	15 freeze-thaw cycles	25 freeze-thaw cycles
Density (kg/m³)	1876	2027	2066	2069
Ultrasonic sound velocity (km/s)	2.063	2.102	1.987	1.837
Compressive strength (MPa)	19.47	8.95	8.28	8.54
Tensile strength (MPa)	2.88	1.31	1.28	1.35
Water content (V%)	1.60	19.62	19.21	19.07

EXPERIMENTAL WEATHERING OF RHYOLITE TUFF 117

pore system. The high effective porosity is important in the selection of conserving methods and in the determination of the penetration depth of the conserving agent. Strength parameters are given by air-dry and water saturated compressive and tensile strength values. The average 19.5 MPa compressive strength of air-dry samples became less than half after water saturation (9 MPa on average). The tensile strength shows similar changes form 2.88 MPa in air-dry samples to 1.31 MPa in the water saturated state indicating that the tuff is sensitive to water (Table 7.2).

RESULTS

Natural weathering and related mineralogical changes

Naturally weathered samples tend to be darker in colour and have a rough surface. Crust formation, scaling and flaking of crusts with rapid material loss are very common decay features of monumental walls and tuff ornaments (Figure 7.4). The texture becomes spongy and tuff tends to crumble. The changes in mineralogical composition between weathered and unaltered tuff were detected by XRD. There is an increase in clay mineral content with the appearance of montmorillonite and chlorite from the slightly altered to weathered tuff samples. Chlorite is the weathering product of biotite. With alteration calcite becomes more and more frequent being the weathering product of Ca-rich groundmass (Figure 7.5). Samples from deeply weathered stone surfaces of Sárospatak Castle have XRD scans with elevated base lines indicating significant amounts of amorphous non-detectable silica, produced by weathering. Since no internal standards were available it was not possible to measure the weathering-related quantitative changes in mineralogy by XRD. Unfortunately, it was not possible to detect mineralogical changes with the above methods on experimentally weathered samples.

118 STONE DECAY ITS CAUSES AND CONTROLS

Figure 7.4 Weathered rhyolite tuff window ledge, Castle of Sárospatak.

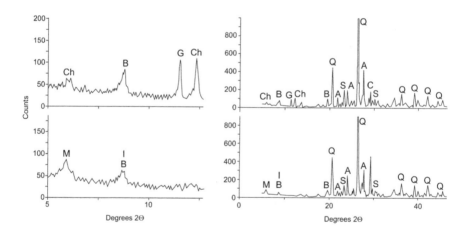

Figure 7.5 XRD graphs of two weathered rhyolite tuff showing the most important XRD peaks of the minerals (A – albite, B – biotite, C – calcite, Ch – chlorite, G – gypsum, I – illite, S – sanidine, M – montmorillonite, Q – quartz). The enlargements of the first parts of the graphs indicate that weathering related minerals such as chlorite (a weathering product from biotite) and other clay minerals such as montmorillonite and illite are also present.

Changes in physical properties by experimental weathering

The standardized freeze-thaw tests (15 and 25 cycles) of 5 cm-high cylinders showed that there is a significant decrease in uniaxial compressive strength, tensile strength and in ultrasound velocity of air-dried and freeze-thaw cycled samples. Following the freeze-thaw cycles both the compressive strength and tensile strength became less than 50% of the air-dry strength (Table 7.2).

Another series of specimens, discs of 2 cm in height, were used for magnesium sulphate crystallization (2 cycles) and for 15 freeze-thaw cycles for comparison. The surfaces of discs have been eroded by frost action and by magnesium sulphate crystallization. These changes were detected by measuring surface micro-roughness. The most drastic effect is caused by Mg-sulphate crystallization when a material loss of a millimetre was measured (Figure 7.6). The surfaces of cylindrical samples became spongy and the edges were rounded (Figure 7.7). The stress caused by crystallization caused the opening of surface parallel to oblique macro- and micro-cracks that are also visible under the microscope (Figure 7.8).

The tensile strength of test samples decreased from 3.13 MPa (air-dry) to 2.82 MPa after 15 freeze-thaw cycles and to 1.80 MPa after

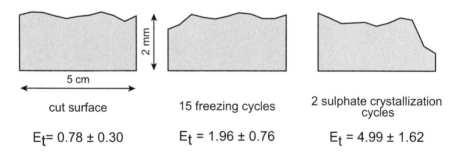

cut surface — 15 freezing cycles — 2 sulphate crystallization cycles

E_t = 0.78 ± 0.30 E_t = 1.96 ± 0.76 E_t = 4.99 ± 1.62

Figure 7.6 Differences in microroughness of cut, frost and sulphate treated surface before PMMA treatment. Et values represents the square of standard deviations from the ideal surface. Note that the vertical scale is enlarged.

120 STONE DECAY ITS CAUSES AND CONTROLS

Figure 7.7 Rhyolite tuff sample discs (5 cm in diameter) after two magnesium sulphate crystallization cycles. Note the salt weathering related rounded edges and the open cracks of the samples.

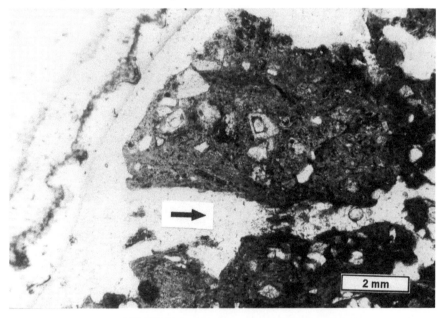

Figure 7.8 Microphotograph of a crack that opened due to sulphate crystallization. The thin-section was prepared perpendicular to the surface of the cylindrical test sample.

Table 7.3 Change of physical properties by experimental weathering of disc-shaped tuff samples (average values).

Properties/Test conditions	Air-dry	Water saturated	15 freeze-thaw cycles	2 MgSO$_4$ cycles
Ultrasonic sound velocity (km/s)	1.959	2.115	1.936	1.485
Tensile strength (MPa)	3.13	1.03	2.82	1.80
Duroscope rebound	22.67	19.79	16.78	8.25

Mg-sulphate crystallization. A similar trend was observed in Duroscope rebound values and in ultrasound velocities (Table 7.3). It is also necessary to stress that water saturation of the tuff has the same effect on strength as frost or magnesium sulphate (Table 7.3). This decrease in strength shows that rhyolite tuff is very sensitive to water absorption and its load bearing capacity decreases not only with frost action or salt crystallization but even with increased water content.

Physical properties of conserving agent (PMMA) treated rhyolite tuff

Water absorption properties were not changed even after using different treatment techniques indicating that the conserving agent is not hydrophobic and thus allows water to penetrate into the rock (Figure 7.9).

Longitudinal ultrasound velocities show an increase after treatment. This suggests that the conserving agent acts as a binder between solid rock particles and strengthens the rock. After additional Mg-sulphate crystallization cycles the sound velocities drop nearly to half indicating the loosening of the texture and a decrease in binding effect of PMMA (Figure 7.10).

Surface strength properties of rhyolite tuff increase significantly after treatment as is shown by Duroscope rebound tests. The

Duroscope rebound values also increase from 8.25 to 13.46–16.08 (PMMA treated after 2 Mg-sulphate crystallization, Figure 7.11) and from 16.78 to 18.38–19.92 (PMMA treated after 5 freeze-thaw cycles) and their standard deviations also decrease suggesting that treated tuff surfaces become more homogeneous. Yet, after further freezing or Mg-sulphate crystallization the Duroscope rebound values show a 50% decrease (Figure 7.11).

Tensile strength test results indicate that neither the surface treatment nor the saturation treatment caused any adverse change in strength. In contrast, the treated samples show increased strength compared to non-treated samples (solid squares on Figure 7.12). The additional 5 freeze-thaw or 2 magnesium sulphate crystallization cycles cause a loss of strength. This weakening is more severe when the surfaces of the samples are only immersed into the PMMA and the test discs are not fully saturated (Figure 7.12).

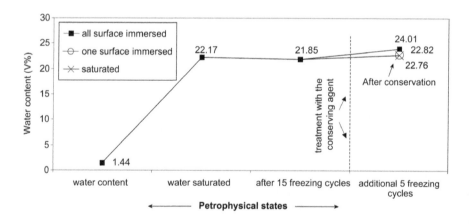

Figure 7.9 Water content of cylindrical samples in different petrophysical states. The PMMA conserving agent is not impermeable and not hydrophobic since the water saturation is nearly the same after 15 freezing cycles, PMMA treatment and additional 5 freezing cycles.

EXPERIMENTAL WEATHERING OF RHYOLITE TUFF   123

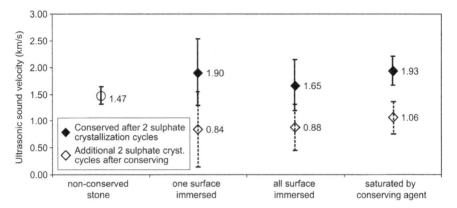

Figure 7.10 Ultrasonic sound velocities after two sulphate crystallization cycles and conserving and additional two sulphate crystallization cycles show that the PMMA agent has only short term advantages by strengthening the rhyolite tuff but after additional sulphate crystallization a significant decrease in ultrasonic sound velocities are observed (vertical lines are the standard deviations).

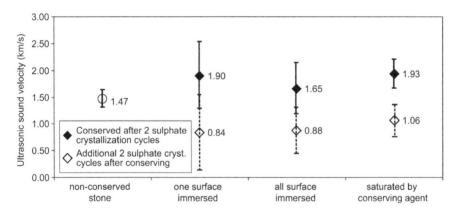

Figure 7.11 Duroscope rebound values after two sulphate crystallization cycles and conserving by PMMA and additional two sulphate crystallization cycles. Note that the Duroscope values show an increase after PMMA treatment but after further crystallization 50% of strength loss marks that the PMMA is ineffective against sulphate crystallization related weathering (vertical lines are the standard deviations).

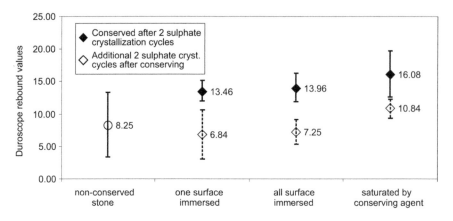

Figure 7.12 Tensile strength after two sulphate crystallization cycles and conserving by PMMA and additional two sulphate crystallization cycles. The strength of conserved samples increases but after 2 additional magnesium sulphate crystallization cycles there is a significant decrease in the strength of surface immersed samples. This type of consolidation models the conserving practice of the monuments (vertical lines are the standard deviations).

DISCUSSION

It was possible to analyse the penetration depth of the PMMA conserving agent when test discs were cut. A darker zone within the tuff showed the penetration depth of the agent. This continuous PMMA saturated zone was 2–4 mm thick when the surface was immersed into the agent for 5 minutes. The 100-hours immersion in PMMA caused the full saturation of void spaces (see Figure 7.2). These differences in treatment are important in the evaluation of test results. The 'one surface immersion' method provides the best analogy to the 'conserving practice', i.e. during restoration works generally the outer surface of the stone (tuff) wall is treated. The 'all surface' immersion for 5 minutes gives information on the behaviour of the PMMA with respect to water repellency. The test results of the fully saturated discs (100-hours saturation) helps understanding of the mechanical properties of PMMA.

The PMMA treated samples show an increased tensile strength, and Duroscope rebound value even when a frost or sulphate crystallization affected irregular surface is treated (Figures 7.11 and 7.12). This indicates that PMMA forms a polygonal internal structure within the porous tuff and acts as a binder. This strengthening remains as long as no further freeze-thaw or sulphate crystallization cycles act on the tuff. If these do occur, the strength of the treated tuff surface significantly drops, i.e. Duroscope rebound values show a 50% decrease after 2 additional magnesium sulphate crystallization cycles (Figure 7.11). The limitations of the agent are also indicated by the very low ultrasound velocities of test discs after two additional magnesium sulphate crystallization cycles (Figure 7.10).

Since high water content has a similar effect on tuff as frost or magnesium sulphate crystallization (see Table 7.2) it is crucial to exclude water to prevent weathering related weakening of the tuff. The PMMA conserving agent is not a water repellent and does not form an impermeable surface on the tuff, as is shown by water adsorption tests (Figure 7.9). Consequently water and salt-rich solutions could penetrate below the treated surface and the crystallization induced pressure lifts up the PMMA-saturated crust, leading to loss of the treated tuff surface. As a consequence the surface treatment by PMMA conserving agent is ineffective against long-term external effects such as freeze-thaw, or salt crystallization and cannot prevent freezing- and crystallization-related structural damage. Thus the strengthening effect of this agent can only be used for interiors, where no water (frost) or salt attack of the tuff structure is expected. It is also necessary to take into account that although PMMA is transparent it also causes a visible surface change, whereby the treated tuff surface becomes darker.

By forming a strong crust on stone surfaces PMMA might be applied to strengthen stone surfaces other than rhyolite tuff. Most probably it can be used on stones where water content does not cause such a drastic drop in strength and where the bond between the treated zone and the host rock is stronger. Sandstone or limestone surfaces can be consolidated with methyl methacrylates,[7] but this example of rhyolite tuff shows that special care and experimental testing are needed before PMMA application.

CONCLUSIONS

The pumice-rich rhyolite tuff of north-eastern Hungary has a high effective porosity (20% volume) and contains weathering sensitive minerals such as biotite and a decayable glassy groundmass. In the course of natural weathering biotite partly transforms to chlorite, clay minerals appear in the tuff and calcite is formed. For the preservation of rhyolite tuff monuments it is essential to impede water penetration into the stone since the strength of water saturated tuff is 50% of that when air-dry. Freeze-thaw cycles and magnesium sulphate crystallization cycles lead to the loosening of the rock texture, an increase in surface roughness and loss of strength.

The tested polymethyl methacrylate consolidant strengthens the rock by forming polymers within voids as is shown by tensile strength and Duroscope rebound values. However, the traditional PMMA treatment, in which the outer stone surface is saturated by the agent, cannot prevent the decay since it only forms a thin load-bearing but permeable crust on the tuff surface. By homogenizing the surface, PMMA temporarily improves surface strength but additional freeze-thaw cycles or salt crystallization lift up this consolidated crust. The short-term advantages of treatment are overshadowed by long-term disadvantages, namely the treated tuff becomes more prone to weathering. As a consequence PMMA can only be applied to consolidate interior, always dry tuff surfaces and not exposed ones.

The application of methyl methacrylate consolidants on other stone types such as sandstone or limestone is more promising especially when the rock is darker, so that no colour change is caused by PMMA, and when the stone is not sensitive to water, so there is no drop in strength on saturation. The example of rhyolite tuff shows that special attention is needed before applying any consolidants.

The above documented petrophysical tests help in the assessment of experimental stone decay and in describing the general effect of conserving agent or consolidant on durability.

Acknowledgements

The critical comments of B. Smith and an anonymous reviewer are very much appreciated and helped in improving the quality of the paper. The study has been partly financially supported by Hungarian National Research Fund (OTKA) and additional financial help was provided to Á. Török by Széchenyi Grant and OM-Mecenatura Fund. Technical help from Gy. Emszt, E.L. Árpás, E. Horthy and E. Saskői is acknowledged.

References

1. McGreevy, J.P. (1996) Pore properties of limestones as controls on salt weathering susceptibility: a case study. In Smith, B.J. and Warke, P.A. (eds.) *Processes of Urban Stone Decay*. Donhead Publishing, London; 150–167.
2. Pavia Santamaria S, Cooper T.P. and Caro Calatayud S. (1996) Characterisation and decay of monumental sandstone in La Rioja, Northern Spain. In Smith, B.J. and Warke, P.A. (eds.) *Processes of Urban Stone Decay*. Donhead Publishing, London; 125–132.
3. Goudie, A.S. and Viles, H. (1997) *Salt Weathering Hazards*. John Wiley and Sons, Chichester.
4. Goudie A.S. (1999) A comparison of the relative resistance of limestones to frost and salt weathering. *Permafrost and Periglacial Processes*, 10: 309–316.
5. Rodriguez-Navarro C. and Doehne E. (1999) Salt weathering: influence of evaporation rate, supersaturation and crystallization pattern. *Earth Surface Processes and Landforms*, 24: 191–209.
6. Ashurst, J. and Ashurst, N. (1988) *Practical Building Conservation*, Volume 1. English Heritage Technical Handbook, Gower Technical Press, Hampshire.
7. Ibid.

8 Chemical Composition of Precipitation in Kraków: its Role in the Salt Weathering of Stone Building Materials

W. WILCZYŃSKA-MICHALIK

ABSTRACT

This study identifies relationships between the chemical composition of atmospheric precipitation and salt weathering processes on stone building materials in Kraków. The composition of precipitation water and concentration of ions varied in different months of the study period. The average monthly influx of different components varies for SO_4^{2-} from 650–2300 mg/m², HCO_3^- from 690–6100 mg/m², Ca^{2+} from 130–1070 mg/m², and Mg^{2+} from 17–470 mg/m². Run-off waters in Kraków are sulphate-rich solutions. In summer the bicarbonate ion concentration is very high. The mineral composition of dry residues after precipitation is dominated by gypsum, but with other sulphates, nitrates, carbonates, phosphates, and chlorides also present.

Surveys indicate that salt weathering in Kraków is especially effective on highly porous stones. Porous sandstones and limestones are subject to intense granular disintegration and scaling. The occurrence of a compact, continuous gypsum crust (and sometimes crust of other composition, e.g. other sulphates, halite) on stone surfaces and within pores is common. During the detachment of gypsum crusts, stone components are peeled from the stone. Salt weathering in Kraków is related mainly to high concentrations of anthropogenic air-pollution typical of urban and industrial areas.

INTRODUCTION

The destructive role of salts has been known since ancient times and salt weathering is important in natural environments of negative water balance (arid zones), in areas of high salt supply from the atmosphere (e.g. close to coasts), or where salts are supplied from groundwater.[1,2] The weathering of building materials of different types (stone, concrete, plasters, bricks) in heavily polluted urban atmospheres, where the influx of numerous salts in rainwater is very high, is caused by the same factors which operate in 'natural' salt weathering.[3] Both wet and dry deposition of salt components is important.

In many urban environments, the main reason why building materials decay is often the reaction between the components of the stone (or other building material) and the atmospheric pollutants, and the crystallization of new components from water held on the surface and in pore-spaces. The newly formed minerals exert a significant pressure on the building material during crystallization.[4,5,6] In addition, the linear thermal expansion of numerous salt minerals, such as gypsum, is significantly higher than the thermal expansion of primary rock-forming minerals, such as calcite.[7] The pressure related to the changes in linear dimensions of salt minerals, which undergo a change in their state of hydration, is also important. In natural salt weathering one can also observe complex reactions because of the numerous interactions between salt components during crystallization and reactions with stone components. For example, the solubility of gypsum can increase when other salts (nitrates, chlorides) are present in the solution.[8] The crystallization of gypsum can also induce chemical reactions of stone components e.g. dissolution of silicate minerals.[9]

The aim of this study is to identify relationships between the chemical composition of atmospheric precipitation and salt weathering processes on stone building materials in Kraków. The usage of diverse types of stone in Kraków creates an opportunity for comparison of their durability in an urban atmosphere and their different responses to ambient conditions. The crystallization of numerous salts inside the superficial layers, continuous dark crusts and efflorescences on stone buildings has been described previously in Kraków.[10,11,12,13,14,15,16,17] Similarly the relation between the crystallization of salts and the chemical composition of atmospheric precipitation was pointed out in a preliminary study.[18] In this study, salt assemblages were found to comprise about twenty minerals, with gypsum being the most common and volumetrically significant. Other assemblages dominated by nitrates occur only locally, mainly in the permanently wet zones of walls. Halite is present in gypsum crusts in negligible amounts but 'monomineral' halite crusts or efflorescences can occur in local environments.[19] In this paper the relationships between these salt assemblages, resultant damage and environmental conditions are examined in detail.

METHODS OF STUDY

Atmospheric precipitation was collected in Kraków in one-month cycles for over one year. Chemical analyses of precipitation water were then performed to determine the content of SO_4, Cl, NO_3, HCO_3, PO_4, Mg, Ca, Na, K, Fe, and Mn. The dry residue after evaporation was studied for determination of 'mineral' composition. The residue was obtained by evaporation in a temperature not exceeding 50°C.

X-ray diffraction and infrared spectrometry was used to identify phases (minerals) in dry residues and in stone samples. The structure of superficial stone layers in thin-sections perpendicular to the surface was examined using optical microscopy. The morphology and chemical composition of primary and secondary components of the zones of stone reaction with the atmosphere were studied by means of a scanning electron microscope equipped with energy dispersive spectrometer.

Stone samples were collected from different buildings in Kraków. Sampling sites were situated high enough to avoid contamination by salts from road, pavements or groundwater. Artificial inputs of chemicals related to conservation, repair and painting were also excluded through careful selection of sites based on knowledge of previous conservation intervention.

CHEMICAL COMPOSITION OF ATMOSPHERIC PRECIPITATION

Samples of atmospheric precipitation were collected in one-month cycles. Concentrations of selected ions in precipitation in Kraków exhibit significant seasonal variations (Table 8.1). The highest concentrations of several components are four times higher than the lowest ones (Table 8.1 and Figure 8.1). For the majority of components the highest concentrations are noted in the winter period. The pH value varies from 5.2–6.8.

Taking into account the volume of atmospheric precipitation per month and the concentration of ions in one-month precipitation samples the influx of ions was calculated. The monthly influx of different chemical components is higher than $1g/m^2$ (Table 8.1 and Figure 8.2). The highest influx of several components does not coincide with the highest concentration in precipitation (e.g. SO_4^{2-} in June 1998). Recalculating the influx of Ca^{2+} and SO_4^{2-} into 'normative' gypsum, which can hypothetically precipitate, it is noticeable that in several months the amount of this component is high (in November 1998 more than 4 g/m^2). The amount of hypothetical gypsum can exceed 3 g/m^2 per month both in wintertime, when coal-based heating is significant, and in summer, when the precipitation is extraordinarily high.

Table 8.1 Concentration of selected components in rainwater and their monthly influx (October 1998–October 1999).

Concentration [mg/l]	Oct. 1998	Nov. 1998	Dec. 1998	Jan. 1999	Feb. 1999	March 1999	April 1999	May 1999	June 1999	July 1999	Aug. 1999	Sept. 1999	Oct. 1999
Cl^-	13.0	8.0	5.0	10.0	6.0	6.0	2.0	2.0	2.0	2.0	2.0	2.0	<1.0
NO_3^-	2.4	2.7	2.3	2.8	2.4	2.0	1.3	1.2	1.4	0.9	1.1	1.8	0.9
HCO_3^-	12.2	18.3	24.2	36.6	30.4	42.7	42.7	42.7	36.6	36.6	18.3	24.6	24.4
SO_4^{2-}	15.8	23.3	56.0	65.4	32.1	59.4	35.3	17.9	13.8	11.5	14.2	18.2	17.3
PO_4^{3-}	0.066	0.475	0.037	0.019	0.015	0.023	0.026	0.05	1.11	0.34	0.053	0.04	0.195
Na^+	0.33	0.46	1.4	0.71	0.5	0.75	0.15	0.05	0.46	0.5	0.23	0.29	0.1
K^+	0.45	0.19	0.37	0.35	0.11	0.45	0.45	0.2	0.6	1.3	0.2	0.25	0.2
Ca^{2+}	8.29	11.1	23.5	29.0	9.7	34.2	14.2	7.1	5.69	2.8	2.79	5.14	2.88
Mg^{2+}	4.1	4.9	6.6	8.3	7.00	9.4	4.3	3.4	3.82	0.36	0.36	0.43	2.15
Monthly influx [mg/m²]													
Cl^-	735.8	769.6	153.0	240.0	177.0	97.2	95.0	86.6	334.0	161.0	92.8	83.6	<80
NO_3^-	135.8	259.7	70.4	67.2	70.8	32.4	61.8	52.0	233.8	72.5	51.0	75.2	72.4
HCO_3^-	690.5	1760.5	740.5	878.4	896.8	691.7	2028.3	1848.9	6112.2	2946.3	849.1	1028.3	1961.8
SO_4^{2-}	894.3	2241.5	1713.6	1569.6	947.0	962.3	1676.8	775.1	2304.6	925.8	658.9	760.8	1390.9
PO_4^{3-}	3.7	45.7	1.1	0.5	0.4	0.4	1.2	2.2	185.4	27.4	2.5	1.7	15.7
Na^+	18.7	44.3	42.8	17.0	14.8	12.2	7.1	2.2	76.8	40.3	10.7	12.1	8.0
K^+	25.5	18.3	11.3	8.4	3.2	7.3	21.4	8.7	100.2	104.7	9.3	10.5	16.1
Ca^{2+}	469.2	1067.8	719.1	696.0	286.2	554.0	674.5	307.4	950.2	225.4	129.5	214.9	231.6
Mg^{2+}	232.1	471.4	202.0	199.2	206.5	152.3	204.3	147.2	637.9	29.0	16.7	18.0	172.9

134 STONE DECAY ITS CAUSES AND CONTROLS

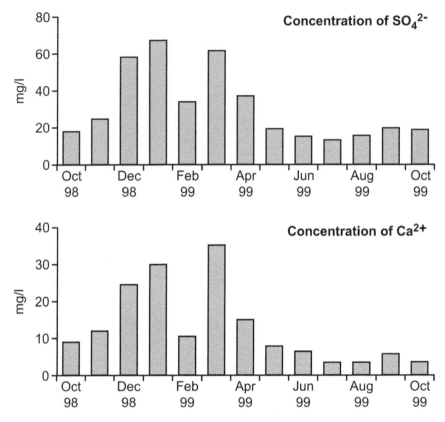

Figure 8.1 Average concentrations of SO_4^{2-} in atmospheric precipitation (October 1998–October 1999).

The phase composition of dry residue after evaporation of precipitation water varies slightly for different months. Gypsum is the major component in all residue samples (Figure 8.3). Besides gypsum other sulphates, nitrates, phosphates, carbonates (calcite, dolomite), and chlorides (halite) are also present in minor amounts. Phosphates (e.g. brushite [$CaPO3(OH).H_2O$]) are present mostly in samples from the summer period. The results of XRD analyses of samples from the winter period suggest the presence of graphite-like substances related to elevated amounts of finely dispersed soot in the atmosphere.

CHEMICAL COMPOSITION OF PRECIPITATION IN KRAKÓW 135

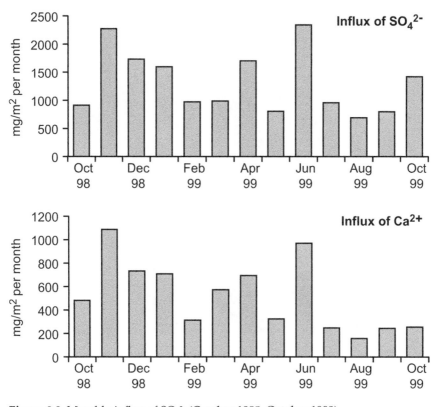

Figure 8.2 Monthly influx of SO_4^{2-} (October 1998–October 1999).

The low content of carbonates in the dry residue after evaporation noted also in samples with HCO_3^-/SO_4^{2-} ratio $c.3$ (often below the detection limit of X-ray diffraction) is caused by the relatively low pH value of water inhibiting carbonate crystallization.

136 STONE DECAY ITS CAUSES AND CONTROLS

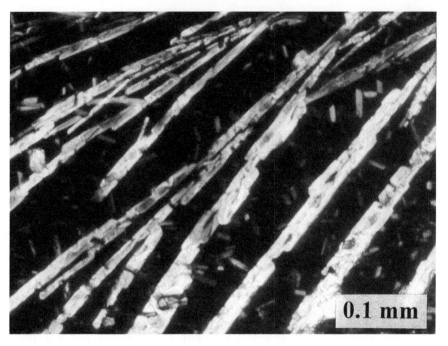

Figure 8.3 Gypsum crystals precipitated from rainwater (optical microscope; crossed polars)

STRUCTURE OF WEATHERING ZONES

Salts in weathering zones of different stones

Gypsum is the dominant newly formed mineral in the zones of stone reaction with the atmosphere in Kraków. Other components identified in these zones (Table 8.2) occur in small amounts. The assemblage comprises other sulphates, nitrates, carbonates and chlorides. The concentration of these components in efflorescences or salt crusts is related mostly to the humidity of the local environment.

Table 8.2 Minerals present in secondary crusts on building stones in Kraków.

Mineral	Chemical formula
Gypsum	$CaSO_4.2H_2O$
Bassanite	$CaSO_4.5H_2O$
Hexahydrite	$MgSO_4.6H_2O$
Epsomite	$MgSO_4.7H_2O$
Melanterite	$FeSO_4.7H_2O$
Langbeinite	$K_2Mg_2[SO_4]_3$
Mirabilite	$Na_2SO_4.10H_2O$
Syngenite	$K_2Ca[SO_4]_2.H_2O$
Burkeite	$2Na_2SO_4.Na_2CO_3$
Calcite	$CaCO_3$
Dolomite	$CaMg[CO_3]_2$
Halite	$NaCl$
Sylvite	KCl
Nitronatrite	$NaNO_3$
Nitrammite	NH_4NO_3

Gypsum in sandstone

Gypsum replaces primary cements (and partly matrix) in sandstone. The final stage of the process is the occurrence of gypsum-cemented sandstone (Figure 8.4). Gypsum-cemented sandstone exhibits a 'float texture' i.e. detrital quartz grains dispersed in gypsum cement. The less advanced stage of gypsum growth in the superficial layers of sandstones is more common. Different size nests of gypsum filling pore-spaces or oversized pores and veins of gypsum are present in numerous types of sandstone.

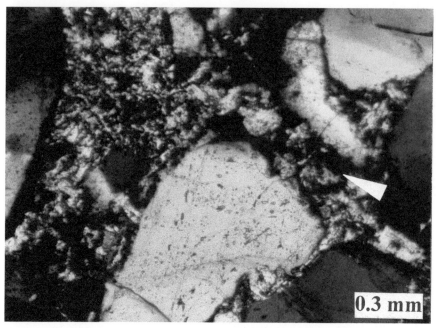

Figure 8.4 Fine-grained gypsum (arrow) totally filling pore spaces in sandstone (optical microscope; crossed polars).

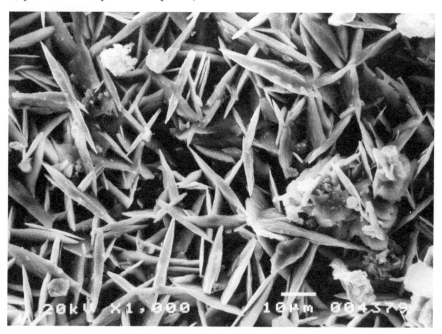

Figure 8.5 Scanning electron micrograph showing euhedral gypsum crystals on sandstone surface (Kraków).

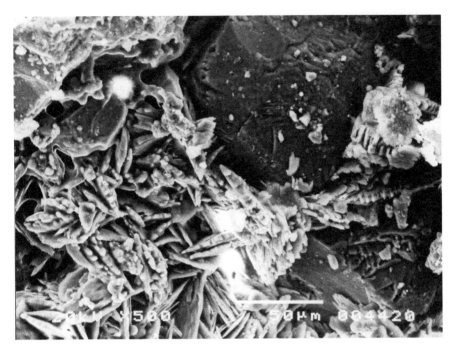

Figure 8.6 Scanning electron micrograph showing platy gypsum crystals growing between quartz grains on the underside of an exfoliating sandstone crust (Kraków).

Gypsum crystals in surface crust are very often euhedral (platy and lenticular forms) (Figure 8.5). Gypsum crystals dispersed between quartz grains inside the sandstone often exhibit plate-like morphologies (Figure 8.6).

Gypsum in carbonate stone

The form of gypsum occurrence (and other salts) is related to the texture of carbonate stone. The porosity of the stone is perhaps the most important factor.

In Upper Jurassic limestone, a relatively homogenous stone of low porosity, gypsum forms a continuous dark layer on the stone surface (a 'black crust') and thin veins within the stone parallel to the stone surface. Gypsum can also be concentrated in nests, which form by filling voids situated close to the stone surface. It is possible to find large empty pores in the outermost part of the stone and gypsum-filled voids in deeper zones in numerous Upper Jurassic limestone samples. It can be noticed that in urban environments the gypsum crystals on the Upper Jurassic limestone exhibit anhedral morphologies. Irregular crystals contain small amounts of Na, Cl, and other elements. Halite as a negligible component can be found in the gypsum crusts on limestones in urban areas. Gypsum on stone surfaces in rural areas forms euhedral crystals and contains less admixed material.

In the Pińczów limestone a relatively high porosity stone, gypsum is present in numerous voids and between the organic remains in the stone. The continuous outer layer of dark-pigmented gypsum is not as common here as in the Upper Jurassic limestone. In some samples one can find more or less continuous veins of gypsum parallel to the surface.

Figure 8.7 Scanning electron micrograph showing alveolae on the surface of the Libiąż dolomite (Kraków).

In the Libiąż dolomite, a stone with medium porosity and relatively inhomogeneous texture, the outermost layer of stone is rich in newly formed gypsum present between dolomite crystals. Gypsum crystals in the stone are often elongated perpendicular to the surface. In some samples gypsum forms irregular crosscutting veins. The volume of gypsum predominates over primary dolomite in numerous cases. The high concentration of dispersed gypsum in the outer part of the stone is typical of overhanging (or partly overhanging) stone surfaces. Beneath vertical surfaces the amount of gypsum inside the stone is usually lower.

On the surface of Libiąż dolomite blocks numerous alveolae are present (Figure 8.7). The intensity of alveolae development is related to the exposure of walls. White efflorescences and powdered salts occur commonly inside alveolae. Magnesium sulphates have been identified in alveolae together with an admixture of gypsum. In relation to ambient humidity, hexahydrite ($MgSO_4.6H_2O$) or epsomite ($MgSO_4.7H_2O$) is a dominant component among powdered salts. Similar salt assemblages occur in widened fissures in the Libiąż dolomite.

Visual manifestation of salts crystallization

The visual manifestations of the effects of salt crystallization on stone surfaces and inside superficial layers depend on stone type. Poorly adhesive deposits of salt are common on stone surfaces. Sometimes these deposits are the reason for alteration of the original stone colour. Homogenous stones of low porosity are often covered by dark gypsum crusts. Gypsum crusts tend to blister and exfoliate (Figure 8.8). The detachment of stone fragments of variable size and the loss of compact stone fragments along fissures (dependent and independent of stone structure) filled with subflorescences, also occurs. Granular disintegration and loss of stone material is typical for porous stones. Salt efflorescences and subflorescences attack the thin (less than one millimetre to several millimetres thick) outer layer of the

Figure 8.8 Blistering and exfoliating gypsum-rich crust on sandstone blocks (Kraków).

stone, resulting in the peeling away of individual grains and small scales. On inhomogeneous stone, morphological changes of the surface due to the formation of alveolar and honeycomb structures can be observed.

DISCUSSION OF RESULTS

Origin of gypsum

Gypsum and other salts that occur in the superficial layers of stone can originate by precipitation from rainwater solutions. Gypsum is also formed by the reaction of carbonate minerals with sulphur-containing components of atmospheric precipitation (both wet and dry) and may sometimes be associated with microbial colonization. The significance of direct precipitation from solution and of production from the reactions of stone components and atmospheric pollutants differs in relation to the pH of precipitation and the concentration of pollutants in rainwater. It has been suggested that in the urban atmosphere in Kraków, precipitation is important because of the high concentration of these pollutants and the near-to-neutral pH of rainwater.[20] This stands in contrast to rural areas, where the concentration in rainwater is lower, as is the average pH value, and reactions between carbonate stones and the sulphur-containing components in the atmosphere are more important.[21] The described difference results in different structures in the reaction zones on carbonate stones in rural and urban areas. It is noteworthy that reduction of emissions of alkaline industrial dusts in Kraków in recent years has lowered the pH of precipitation.[22] The structure of weathering zones on carbonate stones in the Kraków urban area appears to be changing in response to this modification of ambient conditions.

Manecki et al. presented a view that gypsum growth is related exclusively to the reaction of carbonate minerals with H_2SO_4 in precipitation water.[23] High activity of HNO_3 and HF in precipitation waters were also identified by Manecki.[24] The dominance of gypsum in evaporation residues and the occurrence of gypsum crusts on totally carbonate-free sandstones suggests that direct precipitation of gypsum is also an important source of the mineral on the stone surface.

Gypsum in sandstone

Gypsum in the superficial layers of sandstones can originate both from direct precipitation and from the reaction of atmospheric components with unstable carbonate minerals. Regardless of the origin of gypsum, it can exert a significant stress on quartz grains in the stone and produce a loosened 'float texture'. The dark gypsum crusts found on stone surfaces can also lead to much higher thermal expansion of the superficial layer, which could facilitate exfoliation of the gypsum-rich outer layer. Exfoliating gypsum-rich crusts detach sandstone grains or bigger stone fragments. The solution from which gypsum crystallizes can also penetrate deeply into highly porous sandstones and cause granular disintegration. The intensity of granular disintegration is related to the grain-size fraction dominant in the stone. Coarse-grained sandstone is usually subject to more intense disintegration.

Gypsum in carbonate stone

Gypsum in carbonate stone can be of diverse origin. The role of gypsum growth in disintegration can also vary. Highly porous carbonate stone is subject to granular disintegration similar to that on porous sandstones. The rate of stone decay is related to the size of pores.[25,26] The continuous layer of gypsum crust on stone surfaces or lamina/laminae of gypsum inside the stones both facilitate exfoliation. Gypsum in carbonate stone replaces the primary components and also displaces stone fragments. The microscopic study of weathered stone usually does not provide evidence for the distinction of these two types of gypsum (replacive and displacive gypsum). The amount of gypsum present in the stone depends also on the saturation properties of the stone. The higher concentration of gypsum close to horizontal, overhanging surfaces than beneath vertical surfaces suggests that a longer period of solution stagnation is necessary for deeper penetration by the capillary movement. Saturation of the stone during rapid downward movement of solutions on vertical surfaces is low.

POTENTIAL ROLE OF ATMOSPHERIC PRECIPITATION IN KRAKÓW IN SALT WEATHERING

The quantity of dissolved components in precipitation in Kraków is high enough to cause stone decay by salt crystallization. Taking into account the relatively high influx of Ca^{2+} and SO_4^{2-} (as in November

1998) one can assume that about 4 g of gypsum can crystallize per square meter per month. Assuming that the value of porosity is about 5%, a 60-month period is needed for gypsum to fill up all the pore spaces in the outer 2 mm of the stone. In most cases, however, it is not the total amount of the dissolved salts that penetrates into the stone and crystallizes inside it that is crucial as pores do not need to be completely filled for disruption of the stone to begin. In the above calculation only 'hypothetical' gypsum has been considered, but in natural conditions numerous other salts, although only present in small amounts, are active as well.

The occurrence of other salts in gypsum crusts and considerable admixtures of gypsum crystals is typical of urban environments. The various shapes of gypsum crystals may be related to the presence of different ions or organic substances in solutions.[27,28,29]

The presence of numerous salts (e.g. halite) in crusts and underlying stone increases moisture content in the outer layer. The higher moisture content can in turn accelerate decay processes.[30,31,32]

Dark pigmentation of the gypsum crusts and the presence of salts in outer layers cause significant differences between the thermal expansion of this layer and the underlying stone. Differential thermal expansion may, therefore, contribute to the disruption of the stone structure.[33,34,35]

CONCLUSIONS

The detailed study of stone weathered in the urban conditions in Kraków indicates that numerous salts are present in the zone of stone reaction with the ambient atmosphere and atmospheric precipitation. As well as gypsum, some twenty other salts are present in the superficial layers of the stones examined.

Although both the structure of the superficial layers of weathered stones and the form of gypsum occurrence differ strongly in relation to stone type, it is possible to state that in all cases crystallization of salts is important in the destruction of primary stone texture and stone decay. The visual evidence of stone weathering in Kraków is typical of salt weathering.

The composition and concentration of pollutants in atmospheric precipitation indicate that their role in salt weathering of building stones in Kraków can be very important.

Acknowledgements

The study was supported by KBN (Committee for Scientific Research) grant No. 6P04D01714.

References

1 Ferrier, R.C., Jenkins, A. and Elston A.D. (1995) The composition of rime ice as indicator of the quality of winter deposition. *Environmental Pollution* 87: 259–266.
2 Goudie, A. and Viles, H. (1997) *Salt Weathering Hazards*. John Wiley and Sons: 123–149.
3 Fritz, B. and Jeannette, D. (1981) Pétrographie et côntrole géochimique expérimental de transformations superficielles de grés de monuments. *Sciences Géologique Bulletin* 34: 193–208.
4 Franzini, M. (1995) Stones in monuments: natural and anthropogenic deterioration of marble artifacts. *European Journal of Mineralogy* 7: 735–743.
5 Goudie and Viles, *op. cit.* (1997)
6 Winkler, E.M. (1997) *Stone in Architecture. Properties, Durability*. Springer-Verlag Berlin Heildelberg.
7 Goudie and Viles, *op. cit.* (1997)
8 Zehnder K. (1996) Gypsum efflorescence in the zone of rising dump. Monitoring of slow decay processes caused by crystallizing salts on wall paintings. In Riederer, J. (ed.) *Proceedings 8th International Congress on Deterioration and Conservation of Stone*. Ernst and Sohn, Berlin; 1669–1678.
9 Schiavon, N., Chiaviari, G., Schiavon, G. and Fabbri, D. (1995) Nature and decay effects of urban soiling on granitic building stones. *Science of the Total Environment* 167: 87–101.
10 Haber, J., Haber, H., Kozłowski, R., Magiera, J. and Płuska, I. (1988) Air pollution and decay of architectural monuments in the city of Cracow. *Durability of Building Materials* 5: 499–547.

11 Kozłowski R. and Magiera J. (1989)0 Niszczenie wapieni dębnickich i pińczowskich w zabytkach Krakowa (in Polish). *Przewodnik 60 Zjazdu Polskiego Towarzystwa Geologicznego*, Kraków: 204–209.
12 Manecki, A., Chodkiewicz, M. Konopacki, S. (1982) Results of the mineralogical investigations of the dimensions and causes of the destruction of stone elements in Cracow's monumental Buildings. *Zeszyty Naukowe Akademii Górniczo-Hutniczej, Sozologia i Sozotechnika* 17: 35–68.
13 Marszałek, M. (1992) Damage of stone elements in historical buildings of Cracow – mineralogical and chemical characteristics (in Polish, English summary). *Zeszyty Naukowe Akademii Górniczo-Hutniczej, Sozologia i Sozotechnika* 35: 7–34.
14 Wilczyńska-Michalik, W. and Michalik, M. (1991) Mineral composition and structure of crusts on dolomitic building materials in urban atmosphere in Krakow. *Mineralogia Polonica* 22: 69–78.
15 Wilczyńska-Michalik,W. and Michalik, M. (1995) Deterioration of building materials in buildings in Kraków (in Polish, English summary). *Przegląd Geologiczny* 43: 227–235.
16 Wilczyńska-Michalik, W. and Michalik, M. (1998) Differences of the mechanisms of weathering of the Jurassic limestones related to the concentration of air pollution. In Sulovsky, P. and. Zeman, J. (eds.) ENVIWEATH 96, Environmental Aspects of Weathering Processes, *Folia Facultatis Scientiarum Naturalium Universitatis Masaricii Brunnensis, Geologia* 39: 233–239.
17 Michalik, M. and Wilczyńska-Michalik,W. (1998) The influence of air pollution on weathering of building stones in Kraków. In Sulovsky, P. and Zeman, J. (eds.) ENVIWEATH 96, Environmental Aspects of Weathering Processes, *Folia Facultatis Scientiarum Naturalium Universitatis Masaricii Brunnensis, Geologia* 39: 159–167.
18 Wilczyńska-Michalik, W., Pieczara, P., Łatkiewicz, A. and Michalik, M. (2000) Chemical composition of atmospheric precipitation in Kraków as a factor of salt weathering of stony building materials. In Zioło, Z. (ed.) *Działalność człowieka i jego środowisko", Wydawnictwa Naukowe Akademii Pedagogicznej w Krakowie*: 73–91.
19 Wilczyńska-Michalik, W. and Michalik, M. (1993) Solubility controlled mineral zonation in efflorescences and crusts from building walls. *Mineralogia Polonica* 24: 73–87.
20 Michalik and Wilczyńska-Michalik, *op. cit.* (1998)
21 Michalik and Wilczyńska-Michalik, *op. cit.* (1998)
22 Michalik and Wilczyńska-Michalik, *op. cit.* (1998)
23 Manecki *et al, op. cit.* (1982)
24 Manecki *et al, op. cit.* (1982)
25 Haber *et al, op. cit.* (1988)
26 Goudie and Viles, op. cit. (1997)
27 Edinger, S.E. (1973) The growth of gypsum. *Journal of Crystal Growth* 18: 217–224

28 Cody, D.R. (1979) Lenticular gypsum: occurrence in nature, and experimental determinations of effects of soluble green plant material on its formation. *Journal of Sedimentary Petrology* 49: 1015–1028.
29 Franchini-Angela, M. and Rinaudo, C. (1989) Influence of sodium and magnesium on growth morphology of gypsum $CaSO_4.2H_2O$. *Neues Jahrbuch Mineralogie*, Abhandlungen 160: 105–115.
30 Halsey, D.P., Dews, S.J., Mitchell, D.J. and Harris, F.C. (1996) The black soiling of sandstone buildings in the West Midlands, England: regional variations and decay mechanisms. In Smith, B.J. and Warke, P.A. (eds.) *Processes of Urban Stone Decay*, Donhead, London; 53–65.
31 Winkler, *op.cit.* (1997)
32 Halsey, D.P., Mitchell, D.J. and Dews, S.J. (1998) Influence of climatically induced cycles in physical weathering. *Quarterly Journal of Engineering Geology* 31: 359–367.
33 Halsey, D.P., Dews, S.J., Mitchell, D.J. and Harris, F.C. (1996) Influence of aspect upon sandstone weathering: the role of climatic cycles in flaking and scaling. In Riederer, J. (ed.) *Proceedings of 8th International Congress on Deterioration and Conservation of Stone*, Berlin; 849–860.
34 Goudie and Viles, *op. cit.* (1997)
35 Winkler, *op. cit.* (1997)

9 Interpreting Spatial Complexity of Decay Features on a Sandstone Wall: St. Matthew's Church, Belfast

A.V. TURKINGTON and B.J. SMITH

ABSTRACT

A section of one wall of St Matthew's Church, Belfast, was surveyed and the decay features and stone types mapped. Relationships between the dependent variable (decay feature) and independent variables (stone type and elevation) were not significant. Stone blocks that displayed stable surfaces tended to cluster; blocks that exhibited material loss often occurred in isolation. This paper suggests that the spatial complexity of decay features on a plain wall may be a result of temporal variability in the response of sandstone blocks to environmental stresses.

INTRODUCTION

The spatial distribution of decay features receives less attention from stone decay researchers than the rate at which decay proceeds. This may in part be due to the perception that processes are time-dependent not distance-dependent processes.[1] An understanding of the reasons for spatial variability is, however, pivotal to stone decay studies and conservation efforts. The concentration of weathering on particular areas of a building or stone structure must be appreciated by conservators and designers, as it is often the case that the attributes of a structure which are most prized for artistic or historical merit, such as intricate carvings, are those which suffer most from intensive decay.

Mapping of decay forms is viewed as an important tool for conservators and preservationists (see Young, Chapter 2).[2,3] Decay features often display a spatially complex pattern on buildings and monuments, which may be closely related to the influence of structural design on surface environment.[4] Stone decays fundamentally because of the effects of its immediate environment on its material properties, but there is a wide range of decay features that may be produced. It is clear that the extent and distribution of specific features cannot be entirely explained by micro-environmental conditions created by design features. The response of stone to pollution-related stresses is thus not solely a function of exposure conditions, but may also be controlled by stone characteristics, by complex patterns of moisture flux within, and between, stone blocks and even by prior human intervention.

This paper addresses the question of how spatially complex patterns of stone decay can develop across a wall that has experienced, and continues to experience, apparently similar environmental conditions.

ST MATTHEW'S CHURCH

A preliminary investigation of the nature of sandstone decay on a building in Belfast, St. Matthew's Church, was conducted with a view to establishing the range of decay forms which may be identified and the factors that may control their distribution.

Belfast is situated in an embayment on the east coast of Northern Ireland (Figure 9.1) and is characterized by a temperate maritime climate. The city has a long history of highly polluted conditions[5] and regularly experiences episodes of high levels of atmospheric sulphur dioxide and particulates.[6] This derives in part from the configuration of surrounding uplands, which confine the valley of the River Lagan and renders the urban area prone to high concentrations of pollutants beneath temperature inversions during winter anticyclonic conditions.[7]

St. Matthew's Church is situated near the city centre and docks (Figure 9.1), and is constructed of local Scrabo sandstone, with some ornamental Dumfries sandstone. Dumfries sandstone is a medium-grained strongly-bedded red sandstone, dominated by quartz and feldspar, and has been extensively used as a building stone in Scotland and Ireland. Scrabo sandstone was used for construction in and around Belfast in the late nineteenth and early twentieth

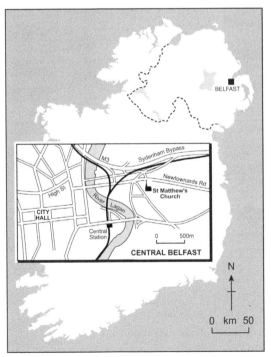

Figure 9.1 Map showing location of Belfast and St Matthew's Church.

centuries.[8] Many of the sandstone buildings in Belfast now exhibit extensive decay, which is frequently characterized by catastrophic levels of material loss. In the case of St. Matthew's Church, by the mid-1990s, granular disintegration, flaking and scaling had progressed to the point where it threatened the structural integrity of the entire building. Decay patterns and rates on individual sandstone blocks on St. Matthew's Church exhibited a high degree of spatial variability across the building, even at a small spatial scale. To examine the causes of this variability, one area of a wall was chosen to identify the nature of decay forms and their spatial distribution (Figure 9.2).

Scrabo sandstone is a Triassic Sandstone, local to the Belfast area. Its detailed composition and structure is variable, but typically it is a clay-rich (smectites) arkosic fine-to-medium grained sandstone, dominated by quartz and potassium feldspar, with trace amounts of muscovite mica. The heterogeneity of Scrabo Stone results from the

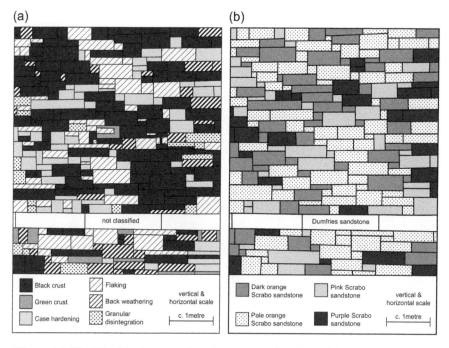

Figure 9.2 Block by block maps of section of east elevation of St. Matthew's Church: (a) decay features, (b) sandstone types.

intrusion of a large Tertiary dolerite sill that is exposed in the major local quarry. The main effects of the intrusive activity were a bleaching of formerly red sandstone to pale pink and buff colours and some hardening of the original friable sandstone.[9] The harder rock, closest in proximity to the sill, was perceived in the nineteenth century to be of good quality and was locally preferred in construction. Consequentially, the aesthetically pleasing variable appearance of Scrabo stone testifies to a wide range of specific characteristics of stone blocks.

As a result of the decayed state of the building, an assessment of the condition of the stone on St. Matthew's Church was conducted by local conservation architects. Through this survey, four types of Scrabo stone were identified (Table 9.1), classified according their performance on the building and described by their physical appearance. Detailed petrographic analysis of these stones was not performed, as collection of large samples from the building surface was prohibited at the time. Apart from occasional stone replacement, the stonework had remained untouched since construction in 1881–1883.

Table 9.1 Classification of Scrabo stone on building.

Scrabo colour	Durability*	Hardness*	Typical types of material loss	Typical appearance on dressing back
Pale orange	Lowest	Lowest	Multiple flaking and sanding	Medium-grained, no obvious bedding
Dark orange	Low	Low	Flaking and scaling	Medium-grained
Pink/yellow	Low	Low	Flaking	Fine-grained
Purple	High	High	Stable surface	Obvious bedding, numerous clay lenses

*as perceived by stone masons/conservation architect

DECAY FEATURES: MAPPING AND SAMPLING

To investigate the distribution of decay features, and relationships between them, one six metre-high section of the east-facing elevation was mapped in detail; the decay forms on each stone block were documented and the stone type noted according to the fourfold classification identified in the architectural survey. The blocks were then classified, on predominance, into three categories of surface hardening and/or deposition and three categories of surface loss. The classification system was adapted from that proposed by Smith et al.,[10] with one extra class, green crust, added to the earlier five-part classification. The classes are: black crusts, green crusts, case hardening, granular disintegration, flaking and back weathering. Black crusts are gypsum-rich; crust thickness and the degree of soiling, due to dust and flyash accumulation, varies considerably. Green crusts are primarily composed of biological material, and appear to be concentrated on Dumfries sandstone blocks and Scrabo sandstone blocks near ground level. Case-hardening represents alteration of the sandstone surface due to reprecipitation of material derived in solution from the stone interior; the dark orange colour is indicative of the surficial concentration of iron oxides. Back weathering, a class used by Fitzner[11] to describe surface retreat, may include multiple scaling, the loss of large (>5 mm) outer layers of stone, which often creates a cavernous form. Flaking is the loss of small (<5 mm) outer stone layers. On St. Matthew's Church, severe material loss is predominantly associated with multiple flaking. Granular disintegration produces a rock meal of salt crystals and loose sand grains, which accumulates at the base of walls. Of the 215 blocks examined, the distribution of the dominant decay features on individual blocks is summarized in Table 9.2 and the spatial distribution depicted in Figure 9.2a. Dumfries sandstone remains unclassified. It is interesting to note that all of the blocks within this area of the wall which were classified as replacement stones at the outset (7 blocks) displayed flaking or back

Table 9.2 Dominant decay types displayed on sandstone blocks on St. Matthew's church.

Decay features	Number of blocks
Black crust	90
Green crust	4
Case hardening	50
Granular disintegration	12
Back weathering	25
Flaking	34

weathering, often with surface retreat of several centimetres. Figure 9.2b shows the distribution of stone types across the section of a wall.

Although classification of entire stone blocks on the basis of their dominant decay feature is simplistic, and decay features on stone may be superimposed, fused or even indistinguishable from each other on any one surface, this classification scheme is easily replicable and transferable to other buildings. Its practicality may, therefore, outweigh any loss of detail incurred through use of a six-part classification. Furthermore, there may be a sound conceptual basis for mapping at the scale of individual blocks. Smith et al.[12] demonstrated experimentally that construction of sandstone walls using lime mortar may cause significant sealing of the sandstone block faces along mortar joints, thus effectively constraining moisture and salt cycling through the exposed surfaces of the blocks.

To examine the influence of stone type on the dominant decay feature, the frequency of each decay feature on each stone type was plotted (Figure 9.3). The frequency of decay features on sections of the wall of various heights above ground was also plotted (Figure 9.4). This was to illustrate the effect of elevation, which may be viewed as a surrogate for environmental variability, on decay distribution. Finally, the 'spatial association', the number of sides of each block adjacent to a block classified in the same decay group, was also mapped and the frequencies plotted in Figure 9.4.

156 STONE DECAY ITS CAUSES AND CONTROLS

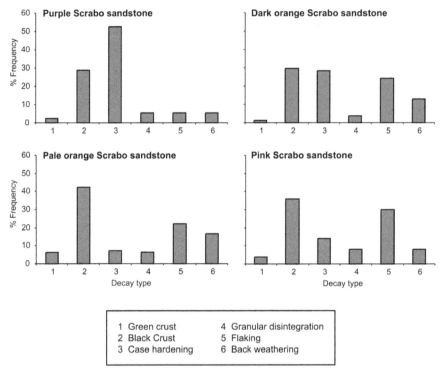

Figure 9.3 Graphs depicting the frequencies of decay features on stones of each type.

SPATIAL DISTRIBUTION OF DECAY FEATURES

Variability in decay features, and in rates of material loss, across the surveyed wall of St Matthew's Church demonstrates that heterogeneity in sandstone properties may be a crucial control on sandstone response to urban environmental stresses. The surveyed sandstone blocks would have been emplaced soon after quarrying and have subsequently been exposed in Belfast in the same location, for the same length of time and have thus experienced the same historical macro-environmental regime. Figure 9.2a illustrates that the spatial pattern of decay appears almost pseudo-random. On a brief visual assessment, it appears that material loss is most severe near ground level, and that blocks displaying granular disintegration and back

INTERPRETING SPATIAL COMPLEXITY OF DECAY FEATURES 157

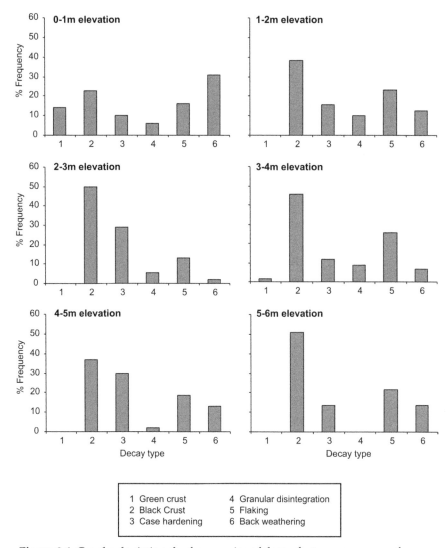

Figure 9.4 Graphs depicting the frequencies of decay features on zones of increasing elevation.

weathering exist in isolation, often surrounded by blocks with stable surfaces. If Figures 9.2a and 9.2b are compared, it would seem that stone type does not control the distribution of decay features, except in the case of the hardest stone type (Table 9.2), which is associated with surfaces that displayed crusts or case hardening. These ideas have been quantitatively assessed, using frequency bar charts to depict relationships between dependent (decay features) and independent (stone type and elevation) variables.

Figure 9.3 depicts the proportion of each decay classification for stones of each of the four types as defined by performance in the building. The 'purple' stone is exceptional in this building as very few of these stones displayed any class of material loss. The other three stone types displayed comparable percentage frequencies of stable surfaces (decay types 1–3) and surfaces actively losing stone material (decay classes 4–6). Although 'purple' Scrabo stone was the least common type, and the most difficult to sample, it did display distinctive characteristics. Primarily, the presence of clays as isolated lenses in a bedded sand matrix. This compares to a more even distribution of clays in the other three stone types. This might suggest that total clay content or the effect of clay content on overall porosity, rather than the expansion of the clay itself, is more significant in determining susceptibility to salt weathering.

Given the nature and exposure of the wall, macro-environmental conditions may be viewed as homogeneous, except perhaps for the influence of elevation. Increasing elevation was expected to be associated with decreasing input of moisture and salts from groundwater. Figure 9.4 shows the frequency distributions for the six decay classes on sections of the wall at 1 m intervals above ground level. The basal 1 m section displayed higher proportions of material loss than the rest of the wall. However, there appears to be no significant difference between the other sections on the basis of the decay types recorded in each. This suggests that the hypothesis of similar environmental regimes across the wall is valid above the basal zone, at least in the context of controls on decay features and their distribution.

Factors controlling stone decay, such as variable uptake of pollutants and salts, differing thermal regime at the exposed stone surface, and variable times of wetness, are ultimately strongly influenced by heterogeneity within the structural and mineralogical properties of the stone. One other factor that may influence the type or severity of decay on a sandstone block is decay of adjacent blocks.

Several surface features may be capable of spreading across mortar joints, especially biological crusts or accretions. Other features may be indirectly affected by an adjacent block, such as the retreat of one block allowing pooling of water and hence accelerated decay of blocks below. The notion that decay features might exhibit spatial clustering was tested using a simple index of 'spatial association', derived from the number of sides of each block in one decay class adjacent to a block classified as the same decay feature (Figure 9.5).

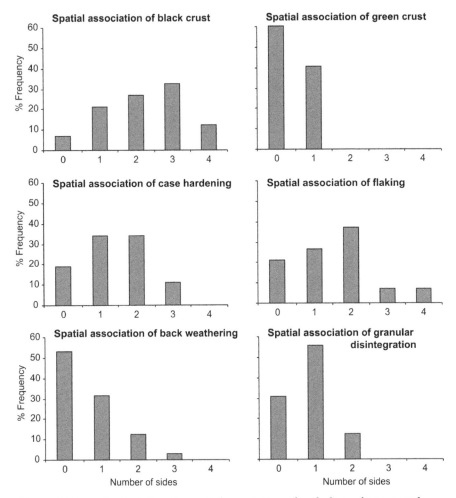

Figure 9.5 Graphs depicting the spatial association of each decay feature, and of stable surfaces and surfaces that exhibited material loss.

Results indicate that blocks that displayed black crust and case hardening tended to cluster together; blocks that displayed green crusts tended to occur singly. Blocks that displayed various types of material loss tended to occur in isolation (Figure 9.5), with more than half of the retreated blocks isolated from any adjacent surface retreat.

INTERPRETATION OF SPATIAL PATTERNS OF DECAY

There are three possible theoretical frameworks for interpreting the spatial clustering of stable, decayed surfaces and isolation of retreating or actively decaying surfaces. This pattern may represent a real relationship between blocks with similarly stable alteration forms, it may arise from the influence of subtle material heterogeneities on thresholds within the stone decay system, or it may be produced by 'heterochronicities' in system development,[13] through which temporal variability in decay begets spatial variability.

Clustering of decay forms of a similar nature may be explained in some cases by actual migration of crust elements across mortar joints; green crusts might offer such an example if biological material colonized adjacent blocks. Similarly, gypsum in black crusts might become mobilized during wetting of stone to be reprecipitated on adjacent surfaces during drying periods. The data presented do not, however, support these ideas and it seems unlikely that there is a true causal relationship represented by clustering of blocks that displayed stable surfaces. Sandstone blocks in a wall are integrated into one structure, but recent research suggests that each block may be viewed as operating as a single small-scale stone decay system, with moisture exchange and salt and pollutant uptake through one exposed face.[14] If this premise is accepted, then the notion of a physical extension of decay forms to neighbouring blocks may be rejected or at least require qualification. Agglomeration of blocks that displayed only surface alteration is likely to be an apparent effect, created by the triggering of rapid retreat in a limited number of sensitive blocks. The fact that substantial sections of the wall did not exhibit rapid retreat is not necessarily an indication that the blocks are not susceptible to retreat, simply that it has not yet begun.

Those isolated blocks which displayed material loss on St Matthew's Church appear to be generally unaffected by the stable blocks surrounding them, and, in turn, have no influence on the progression of decay on adjacent blocks. It appears therefore that susceptibility to material loss is more closely linked with individual stone characteristics. These characteristics may dictate stone response to environmental stresses, and those blocks which have 'failed', in the context of performance of building materials, represent the breaching of a threshold of stability within the stone decay system. Whether this threshold is intrinsic, such as reduced stone cohesion, or extrinsic, such as increased salt supply, is as yet unclear; indeed, the stone decay system is unlikely to be adequately explained in such simple terms. Within the decay system there may be spatial variability in stone response to decay processes at any given time; Schumm termed this phenomenon 'complex response'.[15] Phillips argued that recognition of discontinuous evolution of geomorphic systems and the presence of thresholds is fundamental to surficial process studies.[16] The pattern of decay documented does indicate the heightened susceptibility of some blocks to material loss and surface retreat, presumably as a consequence of specific combinations of material properties.

A third perspective on this issue might be offered by consideration of temporal variability in stone decay as not only integrated with, but creating, spatial variability. An artificial decomposition of time and space should be avoided in stone decay research as differential decay necessarily involves different experiences in locations through time, and locations through time cannot be isolated from adjacent locations. Space, unlike time, is multi-directional;[17] area and distance in any direction play vital roles in the operation of processes through positive and negative feedback mechanisms.[18] However, although time must go forward and feedback controls between decay forms and processes cannot alter past states, a continuity exists between past, present and future states of stone undergoing breakdown in a subaerial environment.

Time, space and their interaction are fundamental concerns in geomorphology. Geomorphologists have long accepted conceptual approaches to problems of process-form interactions, and equilibrium, magnitude and frequency concepts have been extensively reviewed.[19,20,21,22] Some equilibrium concepts, such as dynamic metastable equilibrium, do incorporate thresholds and episodicity in process and response, by including periods of quiescence interspersed by periods of rapid change. The prevalence of nonequilibrium in geomorphic systems is now widely recognized,[23] as is the notion that many systems exhibit multiple equilibria, many of which are unstable. These posits hold promise for stone decay research. Stone decay systems are affected by irreversible processes; once structure is altered, or breakdown is achieved, there can be no reconstitution of the stone. A possible exception to this statement is the removal in solution of precipitated salts or crusts, but stone damage, alteration or material loss, consequent to the presence of such deposited material, cannot be revoked. Equilibrium within decay systems is likely to exist as features which retain their state, or form, within a range of imposed stresses; beyond this range (threshold) the form becomes unstable and a new configuration of the stone is produced, most usually by material loss. Though stone surfaces may appear stable and decay forms may be sustained, deterioration of the stone will continue, particularly in sub-surface or near-surface layers. Stone breakdown may be manifested by intermittent changes in forms, but the trend is continuous throughout the 'lifetime' of the stone or its system. Phillips[24] suggested that equilibrium is more likely to emerge as spatial and temporal scales expand; in the context of stone durability, equilibrium is more likely to be identified as time scales are shortened.

From this perspective, it may be argued that all stones follow a decay pathway, or pathways,[25] and spatial variability in decay features merely represents differing time lags in response by individual stones. The model offered by Smith in which non-calcareous sandstone experiences sequences of case hardening, surface retreat, black crust, surface retreat and so on, is supported by the data presented.[26]

This model, which is reminiscent of metastable equilibrium, incorporates relaxation times and characteristic-form times, as the stone surface reacts to, or internalizes, changes following disruption. Given that heterogeneities in stone properties and in process intensities and duration will result in variable time lags in response of individual stones along this decay pathway, the distribution of decay features across an entire façade will portray increasing spatial complexity as time progresses, a result of heterochronicities, or irregularities, in system development. Mayer demonstrated the importance of specific information regarding dt,[27] the time interval over which a process is realized and causes change, in determining complexities in system behaviour and demonstrated that lags introduce an apparent chaotic behaviour, capable of generating tremendous complexity. In this sense, temporal variations in stone response to decay processes begets spatial variability in decay features.

CONCLUSIONS

This preliminary study has both practical and theoretical implications. The practice of surveying buildings prior to conservation or restoration is certainly valuable. This study indicates the limitations of classifying stone types on the basis of their visual appearance, in particular their colour, as aesthetics have very tenuous links to the material properties exerting control on the nature and rate of decay. It also demonstrates that apparently stable stones may suddenly change, as a result of the breaching of a threshold of stability. This view of decay as an episodic, not linear, phenomenon must be acknowledged by architectural surveyors.

With reference to the stone decay system, the physical and chemical dynamics of decay processes are assumed to be constant, but values of controlling parameters may change over time and space, and positive and negative feedbacks within the system may alter the nature and extent of their control. These controls include environmental parameters and stone properties, and changes to these as decay proceeds. On a plain wall, where environmental parameters are assumed to be relatively invariant, the spatial distribution of decay features cannot be adequately explained by stone type (and therefore properties) or position on the building. The progression of decay through time is not constant, but is affected by stone characteristics, environmental parameters, process characteristics and inherited changes. Temporal variation in stone response to decay, as well as differences in the nature of the response, may create increasing complexity in the spatial pattern of decay observed. The presence and prevalence of thresholds in stone response to decay processes suggests that the stone decay system is, in fact, inherently non-linear. Nonlinearity exists when outputs of energy or material are not proportional to inputs and can lead to complex, chaotic behaviour and patterns.[28,29,30] This is evidenced by the high degree of spatial variability of decay on St Matthew's Church, despite the simplicity of the classification system devised to describe it.

Acknowledgements

The authors are grateful to Gill Alexander, Queen's University Belfast, for cartographic assistance, to Consarc Conservation for providing access to St. Matthew's Church and to John Savage for his patient explanation of the original survey process. The research for this paper was conducted as part of an EPSRC funded project (Grant number GR/L99500).

References

1. de Boer, D.H. (1992) Hierarchies and spatial scale in process geomorphology: a review. *Geomorphology* 4: 303–318.
2. Fitzner, B. (1990) Mapping of natural stone monuments – Documentation of lithotypes and weathering forms. In Proceedings of Advanced Workshop on *Analytical Methodologies for the Investigation of Damaged Stones*, Pavia; 1–24.

3   Fitzner, B., Heinrichs, K. and Kownatzki, R. (1992) Classification and mapping of weathering forms. In Rodrigues, J.D., Henriques, F. and Jeremias, J.T. (eds) Proceedings of the 7th International Congress on *Deterioration and Conservation of Stone* Laboratório Nacional de Engenharia Civil, Portugal, 2: 957–968.
4   Duffy, A.P. and O'Brien, P.F. (1996) A basis for evaluating the durability of new building stone. In Smith, B.J. and Warke, P.A. (eds) *Processes of Urban Stone Decay* Donhead Publishing, London; 253–260.
5   Smith, B.J., Whalley, W.B. and Magee, R.W. (1991) Background and local contributions to acidic deposition and their relative impact on building stone decay: a case study of Northern Ireland. In Longhurst, J.W.S. (ed.) *Acid Deposition: Origins, Impacts and Abatement Strategies*. Springer-Verlag, Berlin; 241–266.
6   Turkington, A.V. (2003, in press) Initial stages of sandstone decay in a polluted urban environment. In Searle, D. (ed.) *Proceedings of SWAPNET 1999*, Wolverhampton, England.
7   Smith *et al.*, *op. cit.* (1991)
8   Ibid.
9   McKinley, J.M., Curran, J.M. and Turkington, A.V. (2001) Gypsum formation in non-calcareous building sandstone: a case study of Scrabo sandstone. *Earth Surface Processes and Landforms* 26: 869–875.
10 Smith, B.J., Magee, R.W. and Whalley W.B. (1994) Breakdown patterns of quartz sandstone in a polluted urban environment, Belfast, Northern Ireland. In Robinson, D.A. and Williams, R.B.G. (eds) *Rock Weathering and Landform Evolution*. Wiley, Chichester; 132–150.
11 Fitzner *op. cit.* (1990)
12 Smith, B.J., Turkington, A.V. and Curran, J.M. (2001) Calcium loading of quartz sandstones during construction: implications for future decay. *Earth Surface Processes and Landforms* 26: 877–883.
13 Trofimov, A.M. and Phillips, J.D. (1992) Theoretical and methodological premises of geomorphological forecasting. In Phillips, J.D. and Renwick, W.H. (eds) Geomorphic Systems. *Geomorphology* 5: 203–211.
14 Smith *et al.*, *op. cit.* (2001)
15 Schumm, S.A. (1979) Geomorphic thresholds: the concept and its application. *Transactions of the Institute of British Geographers* 4: 485–515.
16 Phillips, J.D. (1992) Nonlinear dynamical systems in geomorphology: revolution or evolution. *Geomorphology* 5: 219–229.
17 Thornes, J.B. and Brunsden, D. (1977) *Geomorphology and Time*. Wiley, New York.
18 Thomas, M.F. (1994) *Geomorphology in the Tropics*. Wiley, Chichester.
19 Chorley, R.J. and Kennedy, B.A. (1971) *Physical Geography: a systems approach*. Prentice-Hall, London.
20 Schumm, *op. cit.* (1979)
21 Thorn, C.E. (1988) *Introduction to theoretical geomorphology*. Unwin Hyman, London.

22 Renwick, W.H. (1992) Equilibrium, disequilibrium and non−equilibrium landforms in the landscape. In Phillips, J.D. and Renwick, W.H. (eds) Geomorphic Systems. *Geomorphology* 5: 265–276.
23 Harvey, A.M. (1992) Process interactions, temporal scales and the development of hillslope gully systems: Howgill Fells, northwest England. *Geomorphology* 5: 323–344.
24 Phillips, J.D. (1992) The end of equilibrium? *Geomorphology* 5: 195–201.
25 Smith, B.J. (1996) Scale problems in the interpretation of urban stone decay. In Smith, B.J. and Warke, P.A. (eds) *Processes of Urban Stone Decay*. Donhead Publishing, London; 3–18.
26 Ibid.
27 Mayer, L. (1992) Some comments on equilibrium concepts and geomorphic systems. In Phillips, J.D. and Renwick, W.H. (eds) Geomorphic Systems *Geomorphology* 5: 277–295.
28 Phillips, J.D. (1992) Nonlinear dynamical systems in geomorphology: revolution or evolution. *Geomorphology* 5: 219–229.
29 Mayer, *op. cit.* (1992)
30 Renwick, *op. cit.* (1992)

10 Arkose 'Brownstone' Tombstone Weathering in the Northeastern USA

T.C. MEIERDING

ABSTRACT

Observations on more than 800 Triassic/Jurassic arkose 'brownstone' slab tombstones (1750 to 1850) in 22 cemeteries of the northeastern USA (CT, MA, NY, NJ) demonstrate that those from the northern Connecticut River valley and New Jersey, with fine-grained texture, low mica and high haematite contents, and poorly defined bedding planes are little altered since emplacement. By contrast, coarse-grained brownstones from the southern CT valley are substantially weathered, primarily by contour scaling of a case-hardened surface (up to a mean of 41% surface area lost, Portland, CT). Only in Whippany, NJ could a viable surface erosion rate (0.7 mm/100 years) be obtained, which is clearly lower than erosion rates on other coarse-grained brownstones.

Historic sulphur dioxide input, deduced from measured Vermont marble tombstone surface recession rates at the brownstone cemeteries, is statistically unrelated to amount of brownstone exfoliation. Significantly greater contour scaling on west-facing tombstone faces, together with evidence for groundwater wicking suggests that a non-anthropogenic thermal/moisture process caused the surface breakdown.

INTRODUCTION

Deterioration of outdoor monuments, statuary, and building faces made of carbonate stones has usually been ascribed to urban/sulphur gas atmospheric environments.[1,2] By analogy, air pollution has often been blamed for visually obvious deterioration of other monument and building stones, including the famous arkose 'brownstones'[3,4,5] that were widely used for buildings, monuments, and gravestones in the northeastern USA during the seventeenth to early twentieth centuries. At the time, the stone was considered attractive, easy to quarry, inexpensive, and durable.[6] By the late 1800s, brownstones were considered too sombre for Victorian tastes,[7] and their deterioration on facing stones on buildings in New York City, Boston, Philadelphia, and other cities of the northeast had become obvious.[8]

The primary intent here is to semi-quantitatively (areal face loss percentages) measure brownstone gravestone decay at many northeastern cemeteries to test the tenet that that historic acidic deposition impacted brownstone buildings and monuments. Unlike buildings, with their complicated shapes (e.g. cornices, overhanging roofs, varying foundations), most slab tombstones are simple in shape and in catchment area for precipitation, condensation, air pollution and groundwater uptake. They are rarely impacted by anthropogenic road and sidewalk de-icing salts, or salts/carbonates in building stone mortar. They have little gravity stress from overlying masses. Furthermore, the gravestones are numerous and dated. Most important for this study is that former pollution inputs to the sandstones can be inferred from nearby marble gravestones, which have acted as long-term sulphur dioxide dosimeters.[9,10,11] The northeastern USA is large enough to have cemeteries that have experienced a wide range of sulphur dioxide inputs, but small enough to have a low range of otherwise confounding climatic variations.

BROWNSTONE FORMATION AND DESCRIPTION

Brownstone gravestones in this study were derived from more than 30 quarries within a small region of Mesozoic outcrops (Figure 10.1) in Connecticut[12] and New Jersey[13]. Tilted fault block valleys in

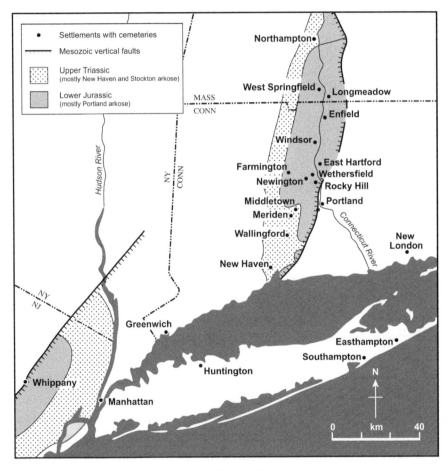

Figure 10.1 Locations of settlements (dots) with cemeteries where brownstone weathering has been assessed in this study. Hatched lines are Mesozoic vertical faults. Lightly stippled land areas are underlain by upper Triassic (mostly New Haven and Stockton arkose) formations, and darker stippled areas are underlain by lower Jurassic (mostly Portland arkose) formations (after Klein, 1968).

Table 10.1 Average volume percent composition from 23 brownstone thin sections (from Hubert, J.F., Gilchrist, J.M., and Reed, A.A. (1982) Brownstones of the Portland Formation. In *Guidebook for Fieldtrips in Connecticut and South Central Massachusetts*. Guidebook 5. State Geological and Natural History Survey of Connecticut: 103–129.)

%	Mineral/rock fragment	%	Cement
26	Quartz	8	Haematite
23	Plagioclase	8	Haematite stains
4	Quartzite	6	Albite
4	Quartz feldspar	2	Clayey matrix
4	Polycrystalline quartz		
4	Schist		
3	Biotite		
2	Muscovite		
2	Schistose quartz		
2	K-Feldspar		
2	Miscellaneous		

New Jersey and in the Connecticut River valley began forming with the break-up of Pangaea at approximately 15°N. palaeo-latitude in the late Triassic. Valley infilling by sandy, gravelly sediments continued from that time (e.g. New Haven arkose, Stockton formation) through the early to middle Jurassic (e.g. Portland arkose). Source rocks for the sediments were gneiss, micaceous schists, and quartzite, and angular fragments of those stone types are seen within the brownstones.[14,15] These sedimentary rocks today display many colours, including tan, grey, black, brown, green, red, and purple, but the brown-red-purple colours were most used as monuments and dimension stone. Red-brown colours are from pink feldspar and from haematite cement.

Hubert *et al.* analysed the composition of 23 brownstone thin sections from three quarries in the Connecticut River Valley

(Table 10.1).[16] None of the clasts or cement are known to be subject to rapid chemical weathering from sulphur gases or acidic precipitation, and the albite cement (precipitated from sodium rich ground water) is assumed to create a structurally tough and durable stone, as does the haematite.

Earlier studies reported that brownstones are also partly cemented by calcite (3–10%),[17,18,19] which would make the sandstone highly susceptible to deterioration under human-altered, acidic atmospheres.[20] Perhaps carbonates in those reports are in stones from underlying Triassic sediments, which are known to have caliche soil layers, because no carbonates were discovered in Jurassic brownstones by Hubert *et al*[21] or in the five thin-sections of tombstones collected for this report, and there was no HCl reaction on the latter five samples, nor from numerous tombstones.

Given the varying Mesozoic climatic, erosional, and depositional environments and later diagenesis conditions, little homogeneity of the quarried stones can be expected. Still, regional generalizations can be made. Excluding conglomerates, the 26 mapped quarries in the Connecticut River valley tend to be spatially clustered into three groups,[22] and thin-section analyses from three quarries and several tombstones demonstrate that composition and grain size vary systematically from north to south. Four northern arkoses (e.g. Redstone Lake quarry, East Longmeadow, MA), have few lithic fragments, are fine to very fine-grained and contain 20–27% haematite stains and cement, which, in the extreme, contributes a brick red colouring ('redstone' in the building trade). By contrast, sandstones from the southerly part of the valley (e.g. Brazos quarry of Portland, CT) are medium to coarse-grained, poorly sorted angular to subangular, lithic arkoses. Volume percent haematite stains and cement is 12–16%, and rock colours are more yellowish brown and purple than in the northern population. Brownstones from the central part of the valley (e.g. Buckland quarry, Manchester, CT, near East Hartford) are intermediate in characteristics between the northern and southern extremes.

Brownstones at four cemeteries evaluated here (Whippany, NJ, Manhattan, Southampton, Easthampton, NY) probably were quarried in New Jersey. Arkoses of New Jersey are similar to northern Connecticut River valley stones in their fine-grained texture,[23] but their brown colour (low haematite content) is more similar to Portland brownstones. They are said to have less visible bedding than the Connecticut River valley brownstones.[24]

Field observations (especially of mica grains), as well as numerous reports on quarrying operations make it clear that all gravestone faces were cut parallel to the bedding. Inscribed faces were usually polished, and back-sides were sometimes polished, sometimes rough.

PREVIOUS ASSESSMENTS OF BROWNSTONE DETERIORATION

Brownstone degradation in Manhattan was noticed early,[25,26] but compared to carbonate stones, measurements of the damage in the outdoor environment are few. Matthias calculated a maximum face recession rate of 1.1 mm/100 years for 66 Portland arkose tombstones in Middletown, CT (cemeteries unknown) by measuring inscription depths for the number '1' in the death dates, and he assumed all had been carved to the same depth initially.[27] He suggested an increase in apparent weathering rate over time, due either to variations in inscription carving depth or rock type over time, or more calcite (then considered to be a brownstone cement) dissolution due to increasing carbonic acidic precipitation from atmospheric CO_2 increases. Rahn applied a descriptive 6-class ordinal legibility scale to tombstone inscriptions on four stone types in a Connecticut cemetery and found the relative order of durability to be granite>schist>marble>Portland arkose sandstone (presumably 25 stones for the latter).[28] Measurement methodologies of Matthias and Rahn were not used here because they are not suited to the observed weathering mechanisms of the brownstones.

ARKOSE SANDSTONE WEATHERING SEQUENCE

Like various other previously described tombstones,[29,30,31,32,33] brownstone surfaces and inscriptions appear fresh (little granular weathering) until large portions of the face fall off as a >3 mm-thick layer. This threshold mode of weathering, (named contour scaling here, spalling or exfoliation in some reports) probably explains why there have previously been so few quantitative surface recession rate measurements on sandstone monuments. Brownstone deterioration consists of at least four steps, each of which could have a different causative environmental and weathering mechanism.

Case hardening

The indurated surface (case hardened) outer layer of brownstones is likely to have formed from precipitation of secondary minerals within pores at the stone surface during evaporation events, with consequent pore/grain cementation and reduction of surface layer porosity and permeability.[34,35,36,37] In the case of tombstones, the source of the precipitated chemicals could be dust, ions and solids in rain, or dissolved ions brought into the stone by capillary rise of soil water. However, it is more likely that precipitated ions dissolved from minerals within the stone (silica and iron in studies by Robinson and Williams)[38], because the stone core is left soft and porous as a result of ion removal. The water that brings the dissolved ions to the surface during evaporation events could be quarry water,[39] rain water,[40] or soil water.[41]

Two coarse-grained brownstone tombstone pieces were picked from small refuse piles in Wallingford, CT and Easthampton, NY. Thin section and SEM transects traversed the stone from face to interior. Unlike in previous studies,[42,43,44,45] no obvious, unequivocal differences can be seen in composition of either matrix or cement between the case-hardened layer and the stone interior under the crack that separates the two, but these samples are too limited to be definitive. Repeated EDAX analysis across the surfaces showed oxide peaks of Si, Al, and Fe. Lesser peaks were Na, K, Mg, and Ca. No sulphur was evident anywhere.

The case hardened layer can form rapidly, and almost certainly formed on tombstones before the major air pollution era (the 1900s). Merrill noted that fresh cut rock at the quarry has so much pore water that it is soft and easy to work, but that within months of being left outdoors it develops a tough surface crust.[46] Indeed, it was common to 'season' monumental sandstone for several years before use to prevent frost burst and to harden the surface. Laboratory studies by Paraguassu[47] and Whalley[48] show that silica from the stone interior can be dissolved and reprecipitated as amorphous silica at the surface very quickly. In the present study, a number of remnant stubs of brownstone slabs that had been broken off close to the ground (presumably by vandalism, time unknown) possess upper surfaces that are hard when struck lightly with a hammer, rather than the soft cores observed immediately after loss of case hardened faces. Therefore, initial quarry water is not the only agent of ion migration to the surface. Williams and Robinson report that even freshly exposed stone interiors can occasionally crust over with silica and iron.[49]

If it can be assumed here that the thickness of the case hardened layer is demonstrated by the surface-parallel crack that develops later, it is unlikely that capillary rise of water from the soil is responsible for its formation. The face slab defined by the crack has a strikingly similar thickness from the gravestone base to the top at 2 m, and in some cases even continues as much as 4 m above the ground on sandstone obelisks. Even though it has been shown that capillary water can theoretically reach 10–30 m above a ground-level source in buildings,[50] it is unlikely that evaporation of capillary water or deposition of its dissolved ions would be so even over such a vertical distance, especially on thin sandstone slabs with their high surface area/volume ratios. Furthermore, case hardened layers form even on stones with low-porosity bases.

Formation of face parallel cracks

Sometime after the case hardened layer is formed, a bedding-plane tensional crack (seen at stone edges) appears at the abrupt, weak interface between the hard layer and the softer core underneath. Case hardened stone surfaces perpendicular to the bedding planes do not exhibit such a face-parallel crack, as shown by brownstone obelisks, which are square in map view, but on which cracks open only parallel to the weak bedding planes and not across them. Where obelisk surfaces slope slightly inward with height, the crack is not precisely parallel to the surface, but parallel to the bedding, with small step-like cracks creating an overall impression of surface crack parallelism. Therefore, the low tensile strength imparted by some aspect of the bedding aids in cracking, as has frequently been noted.[51]

Figure 10.2 Portland arkose gravestone in Middletown, CT (St. John's Catholic Church) has lost most of its thick case hardened surface, exposing a softer core underneath. Although this stone architecture with pedestal was used for a case hardened layer thickness study, only slab tombstones inserted directly in the ground were used in the rest of the study.

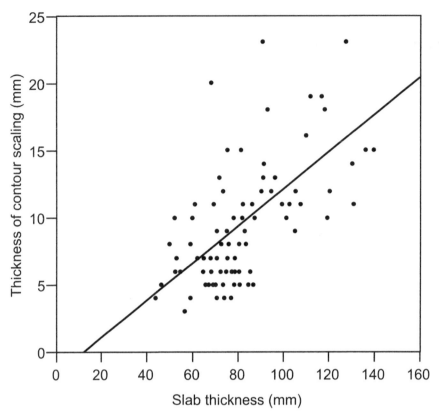

Figure 10.3 Positive relationship between contour scale layer thickness and slab thickness for 89 Portland arkose gravestones in Middletown and Portland, CT. (Best-fit line: y=-1.01+0.13x)

In general, the thicker the Portland brownstone slab, the thicker the hard layer above the crack. Ruler measurement (±1 mm) of the thickness of both the layer on the inscribed face, and corresponding gravestone slab thickness for 89 gravestones of widely varying thickness in cemeteries of Middletown and Portland, CT (Figure 10.2) has a correlation coefficient (r) of 0.62 (Figure 10.3). If the face-parallel, contour-scaled layer can be assumed to be the same thickness and quality as the case hardened layer (not yet established), it may be that thicker stones have more rain and/or groundwater catchment area and more interior surface to dissolve, thus creating a thicker case hardened layer from more water and more precipitable ions.

Figure 10.4 Uneven 'bubbling' of the 3 mm-thick case hardened surface of a stone in Southampton, NY.

Whereas the case hardened layer is strikingly even in thickness, thin layers are commonly bent and bubbled (Figure 10.4). This could be due either to expansion that causes cracking occurring more in some places on the stone face than others or simply due to randomly distributed plastic behaviour of the case hardened layer after separation.

Contour scaling and removal of case hardened layers

Regardless of how the case hardened surface layer is formed and separated from the core, its actual threshold loss from the stone is most meaningful for human perception of monument degradation, and is accordingly the attribute measured in the present reconnaissance study. Potential processes are discussed later.

Weathering of stone interiors

On brownstones, contour scaling exposes softer core rock, which erodes by flaking or granular loss.[52] Those losses were only measured and assessed in Whippany, NJ.

SAMPLING PLAN

Cemeteries in each town founded before 1830 were located on topographic maps. Data were collected from any cemetery that had a large number of brownstones (preferably more than 30). More cemeteries were visited than the 22 included here, but many had few or no arkose tombstones, or the stones had been repaired (e.g. Hartford, CT). Few cemeteries outside the brownstone quarry regions used arkose, except for Long Island, where ship transport was convenient. It is estimated from the search success ratio that there are probably another 10–15 undiscovered cemeteries in the northeastern USA that would fit the requirements for sampling.

Within a cemetery, data were collected by systematically moving along lines of tombstones, beginning at an arbitrarily chosen location. While not a random sample, there is no visual biasing of sample choice. All stones were included if they were slabs inserted directly in the ground, letters were professionally inscribed, and stone faces were relatively polished. The latter requirement excludes some of the earliest arkoses. Most brownstones in this study had death dates from 1750 to 1850 (Figure 10.5), and there is no reason to think that stone emplacement date differs significantly from death date.

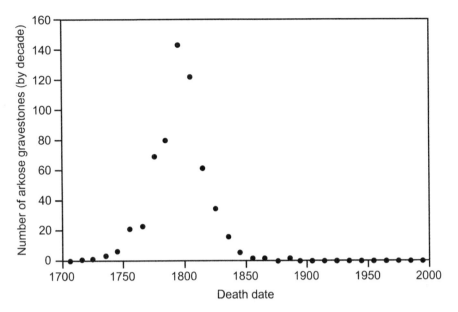

Figure 10.5 Inscribed death dates and number of brownstone gravestones (per decade) measured in this study.

SURFACE RECESSION RATE FROM DEPTH PROBE IN WHIPPANY, NJ

A successful attempt was made to measure above ground surface recession (volume loss) at many sites on many stones at one geographic location (Whippany, NJ), where it was facilitated by the small amount of contour scaling on the fine-grained brownstones. A stiff 1 cm thick plexiglass panel was held by bungee cord against the remaining polished case hardened inscribed face for 30 stones (Figure 10.6). On each stone, a digital depth probe (1×3 mm contact surface) was inserted through drilled holes (10×10 cm grid) in the plexiglass to measure how much the entire above ground sandstone face had receded at each sampling spot, by ignoring inscriptions and subtracting the plexiglass thickness. The amount of surface recession divided by exposure time was averaged for each of perhaps 50 locations on 30 stones, producing a grand mean surface recession rate of

Figure 10.6 Plexiglass plate attached to remaining surface brownstone tombstone in Southampton, NY. Percent face loss due to contour scaling can be estimated within each 10×10 cm grid cell. Holes in centre of grid cells allow depth probe measurements for volume loss.

0.7 mm/100 years (undoubtedly less than in many other cemeteries in this study). Application of a linear time function to threshold surface loss, under varying environmental conditions is necessarily inaccurate, but is reported here for relative comparison of arkose stone erosion rates with other published rates.

AREAL PERCENT SURFACE LOSS MEASUREMENT METHODS

Measurement of absolute surface recession was not possible at most other cemeteries of the study area. Bulges and ripples in the case hardened layer on some stones (Figure 10.4), and complete loss of the surface on others precluded establishing a meaningful pre-weathering reference surface. However, the Whippany experiment confirmed that a more easily estimated, but only semi-quantitative index of surface recession (percent of surface area lost by contour scaling) is spatially correlated with measured surface recession (Figure 10.7), as it should be because contour scaling accounts for most surface recession on stones in this study.

Areal percent surface lost by contour scaling is clearly defined visually, and it was recorded in two ways. To determine the locus of contour scaling on stone faces, a clear plexiglass sheet gridded with marked 10×10 cm squares was placed against the inscribed faces of approximately 30 stones per cemetery (Figure 10.6). The percentage of the original face lost was estimated within each grid square and averaged at each grid square location for the 30 or so stones of a cemetery. Grid squares were only included if at least 20 stones contributed to the average. Some stones were larger than others, and if contour scaling amounts differ between face centres and edges for physical reasons, the differing stone sizes and grid square averaging method could create spurious edge effects. This is not believed to be a major problem for the cemeteries included here. No meaningful purpose would be served by computing contour scaling rates, but given the approximately equal exposure time of the arkoses, relative contour scaling percentage comparisons between cemeteries are reasonably made. Even though the grid square method is simple, it is relatively time consuming to apply, so it was used in only 10 cemeteries (Figure 10.8).

Logistical demands of this extensive geographic reconnaissance dictated use of a quicker and 'dirtier' surface loss measure. Thus, in 12 cemeteries, an estimate was made of percent surface lost by exfoliation for each headstone as a whole, and a mean percent lost calculated for each cemetery and for each of three classified grain sizes (Table 10.2).

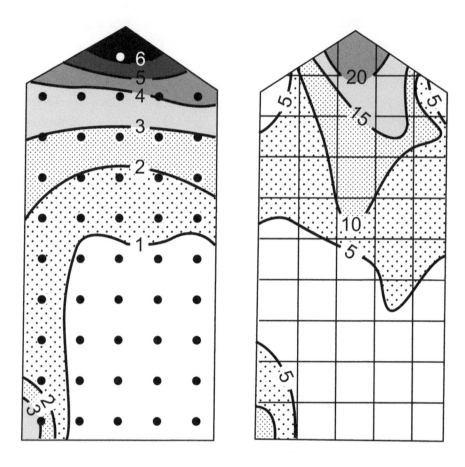

Figure 10.7 Comparison of average amount of surface loss on 30 fine-grained brownstones in Whippany, NJ, by two different methods both involving data collection on a 10×10 cm grid. On the left is mean surface recession rate (mm/100 years) measured with a depth gauge at 46 sampling sites (dots). On the right is mean percent area of surface loss due to contour scaling, by grid cell (grid centroids providing isoline control are not shown). Stone shape and size is generalized.

ARKOSE 'BROWNSTONE' TOMBSTONE WEATHERING 183

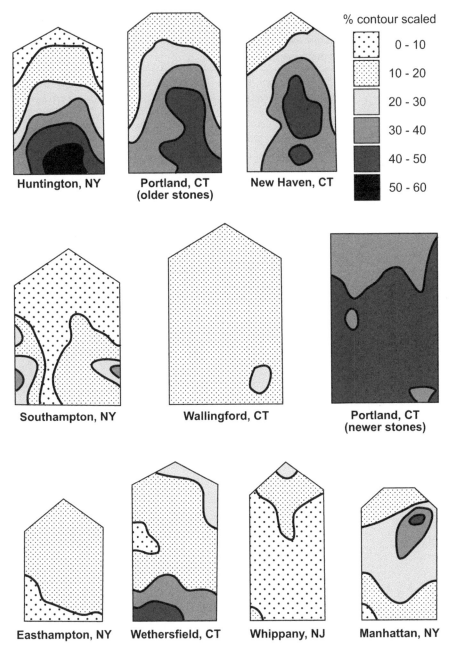

Figure 10.8 Mean areal percent loss of brownstone case hardened layer at cemeteries in the northeastern U.S. Isoline control points are from centres of 10×10 cm grid cells, and grid cell averages were calculated from 20–30 coarse-grained stones per cemetery. Stone shapes and sizes are generalized.

184 STONE DECAY ITS CAUSES AND CONTROLS

Table 10.2 Summary data on tombstone weathering, with towns listed from North to South. Most marble samples are medium-grained Vermont marbles; but starred (*) samples are coarse grained.

Town	No. of sandstone samples	Sand-stone mean date	Sandstone mean % area contour scaled	Sandstone texture (grain size)	Mean marble surface recess. rate (mm/100 years)
Northampton, MA	30	1809	0	Fine	1.15
West Springfield, MA	36	1809	1.1	Fine	0.74*
Longmeadow, MA	30	1797	0	Fine	0.91
Enfield, CT	30	1780	0	Fine	0.6
Windsor, CT	30	1799	3.3	Fine	0.5
East Hartford, CT	60	1811	14.9	Coarse	0.92
Farmington, CT	10	1799	2.2	Fine	0.73*
	9	1781	16.7	Medium	0.73*
	24	1783	36.5	Coarse	0.73*
Wethersfield, CT	179	1806	31.9	Coarse	0.86
Newington, CT	6	1774	3.7	Medium	0.65
	33	1782	18.7	Coarse	0.65
Rocky Hill, CT	40	1798	26.1	Coarse	0.69
Portland, CT	29	1753	29.1	Coarse	0.69
	19	1842	41.1	Coarse	0.69
Middletown, CT		1857		Coarse	0.69
Meriden, CT	12	1788	5	Fine	1.1
	12	1786	5.8	Medium	1.1
	17	1801	22.3	Coarse	1.1
Wallingford, CT	30	1782	15	Coarse	0.96
New Haven, CT	11	1777	0	Fine	0.46
	7	1791	12.9	Medium	0.46
	22	1791	37.3	Coarse	0.46
New London, CT	30	1792	30	Coarse	0.5
Greenwich, CT	21	1797	11.9	Fine	0.11
Easthampton, NY	12	1807	14	Coarse	0.3
Southampton, NY	59	1776	12.1	Coarse	0.29
Huntington, NY	19	1806	5	Medium	0.55
	17	1791	15	Coarse	0.55
Manhattan, NY	20	1805	21	Coarse	2.40*
Whippany, NJ	30	1783	6.3	Fine	0.78

BROWNSTONE PHYSICAL PROPERTIES AND CONTOUR SCALING

Physical properties of the brownstone tombstones have not been quantitatively assessed because non-destructive protocols make it difficult to obtain samples. Even so, it is clear from visual observations that contour scaling areal percentages are correlated with cement type, degree of lamination, and permeability. Unfortunately these properties generally co-vary geographically in the Connecticut River Valley, so their individual influences on stone decay are not easily separated.

Munsell colours were recorded for many tombstone exteriors and interiors as a haematite content surrogate, but they are not reported here because they are entirely too variable. Colours depend not only on stone composition, but also on degree of weathering,[53] soot from historic pollution, organic growths, and occasional white efflorescence. However, it is obvious that very reddish, haematite-rich tombstones from the northern end of the valley (Table 10.2) exhibit almost no contour scaling, even on coarse-grained tombstones. Whereas Geike suggested that iron oxides accelerate sandstone gravestone surface loss rates in Edinburgh,[54] haematite has been shown to protect sandstones in Petra, Jordan.[55] High haematite clay cements contents could help to protect brownstones from contour scaling because of their resistance to chemical change, their occupation of pore space, and their plastic adaptation to mineral or stone expansion.

Merrill noticed that brownstones with less laminated beds exhibit less 'scaling'.[56] This is also evident today in the relatively low amount of contour scaling for less visibly laminated brownstones in Whippany, Easthampton, Southampton, Manhattan, Northampton, Longmeadow, and Windsor. Thin section and field observations show that bedding in brownstones is almost completely defined by mica grains, and that there is much more mica in brownstones from the southern end of the Connecticut River Valley, where the amount of contour scaling is greatest. High biotite contents there could be due to the Boulton micaceous schist source on the eastern border fault in the southern part of the valley,[57] but less so in the north (gneissic source). Alternatively, biotite that might have formerly been in northern brownstones could have been partially converted during diagenesis into the presently abundant haematite cement.[58]

Biotite is known to expand upon weathering, but over longer time spans than the exposure time of brownstone monuments,[59] unless biotite weathering is accelerated by anthropogenic acidic conditions. The few thin sections of Portland arkose from contour scaled tombstones generated here do not demonstrate advanced biotite weathering, and there are no visible expansion cracks near biotite grains. It is more likely that mica grains passively contribute to low rock tensile strength in the bed-perpendicular direction, aiding other contour scaling processes.

Whereas total porosity, pore size, and permeability are known to be important in stone weathering,[60] no data were collected on those properties. Pores are difficult to see and to measure in the few thin-sections made. With little *a priori* justification, it is assumed here that grain size is a crude surrogate for pore size and rock permeability, but not necessarily for porosity because a fine-grained redstone quarry sample has a similar density ($c.2.1$ g/cm) to a coarse-grained Portland brownstone quarry sample.

Following the lead of Hubert *et al*,[61] a crude visual field estimate of the sizes of the larger grains was recorded for each tombstone measured in this reconnaissance study, using the following classes: very fine (<0.1 mm diameter), fine (0.1–0.2 mm), medium (0.2–0.4 mm), coarse (0.4–0.6 mm), and very coarse (>0.6 mm) (Table 10.2). Stones with very fine and very coarse grains were too few to report. Stone ages are very similar between the three crudely defined texture categories.

Even from the poorly defined textural definitions, it is clear that grain size (probably because of attendant permeability variation) is highly inversely correlated with contour scaling of the brownstones. When data from all 22 cemeteries (Table 10.2) are combined, mean total areal percent contour scaling is 25% for coarse-grained stones (n = 591), 12% for medium-grained (n = 35), and 3% for fine-grained (n = 232). Although not measured, very fine-grained stones (most common in northern Connecticut River valley cemeteries) do not contour scale at all, and they show little granular weathering, appearing dark, but fresh.

Grain size is easily categorized, and coarse-grained stones contour scale most. Therefore, only coarse-grained stones are included in further assessing the brownstone contour scaling processes.

STONE EXPOSURE PERIOD AND CONTOUR SCALING

Under unvarying atmospheric and rock conditions, old coarse-grained brownstones should be more deteriorated than young, but that is not quite the case here. Contour scaling areal percent was compared to stone emplacement age for all coarse-grained stones in this study whose dates could be read (Table 10.3). In spite of some bias (dates cannot be read on the most weathered stones, and severely damaged stones have probably been removed), older stones are in better condition. At least in Hartford, CT, Hosley and Holcombe believed that resistant tombstones from Windsor, CT were used earlier, and weaker Portland brownstones later.[62] Massive quarries and historic reports[63] support the view that the later-quarried weak, but more heavily marketed Portland brownstones began to be

Table 10.3 Contour scaling areal percentage and age of all coarse-grained brownstones in the study whose death dates could be read.

Stone Age	Number	Mean % face lost
1740–1759	14	5.3
1760–1779	24	17.4
1780–1799	83	20.9
1800–1819	114	24.0
1820–1839	27	25.7

used in greater numbers than more durable arkoses from New Jersey and the northern Connecticut River Valley beginning approximately 1770, and they continued to dominate the stone building market into the late 1800s. While this age/rock strength trend introduces some bias into the analysis of coarse-grained brownstone weathering processes, it is estimated that 70–80% of the coarse-grained stones in this study are from Portland.

ENVIRONMENTAL INFLUENCES ON BROWNSTONE WEATHERING

Historical sulphur dioxide concentrations

Air pollution in New York was lamented in the late 1800s,[64] but there were no early measurements of sulphur dioxide concentrations. A model that incorporates population, fuel type, industrial activity, and so on,[65] estimates that ambient sulphur dioxide concentrations in cities of the northeastern USA were low in the 1880s and high during the 1900s (peaking around 1940). Measured data from the later 1900s generally verify Lipfert's model results.

Surface recession rates of Vermont marble tombstones, measured by differences in caliper thickness difference measurements between 50 cm height (high surface erosion) versus ground-level (low to no weathering) have acted as a generalized dosimeter of historic SO_2 concentrations at many locations throughout North America[66] and, in fact, correlate highly with levels predicted by Lipfert's model.[67] Therefore, surface recession rates were measured and averaged (n = 30/cemetery) here for Vermont marble vertical tombstones in the same cemeteries as the brownstones (Table 10.2) in order to provide a relative, but at least somewhat quantitative input of average SO_2 concentrations over the last century.

Unfortunately, in several locations (e.g. St. Paul's Cemetery, Manhattan), marble tombstones are coarse-grained (source either Pennsylvania or Massachusetts). Surface recession rates are not identical between Vermont marble and coarse-grained marble tombstones within the same cemeteries above 1 mm/100 years, because coarse-grained marbles contour scale (threshold increase in surface loss) in high sulphur dioxide environments,[68] and medium-grained Vermont marbles do not. Even though marble tombstone surface recession does not proportionately relate to the amount of pollution input in Manhattan as it has for other cities, clearly the former SO_2 concentrations were high there.[69]

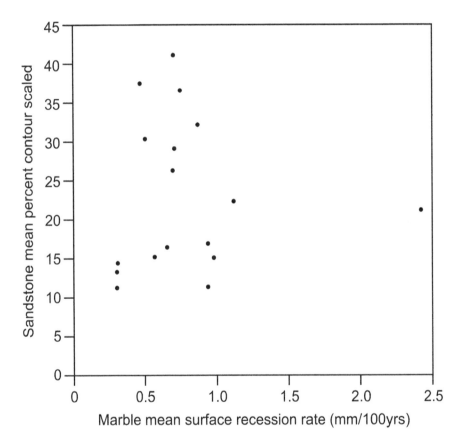

Figure 10.9 No correlation exists between mean Vermont marble recession rates (a SO_2 surrogate) and coarse-grained Portland arkose tombstone mean percent surface contour scaling, both within the same cemeteries (dots) in the northeastern U.S.

There is no correlation (Pearson's r = 0.02) between the mean weathering rates of the sulphur dioxide-affected Vermont marbles and the mean percent contour scaling of the coarse-grained Portland arkoses (Figure 10.9). Sulphur dioxide does not therefore appear to be responsible for their deterioration, which is not surprising given the evident lack of sulphur-sensitive minerals in brownstones. This conclusion is partially supported by historical observations made very early in the pollution era that: 'The [already] cracked and scaling fronts of brownstone in New York and other of our older cities furnish again abundant illustration of lack of care and judgment in the selection of materials...'.[70] Similar statements by Julien show that brownstone weathering was well advanced before SO_2, anthropogenic acid rain and de-icing salts were prevalent.[71]

Groundwater wicking

Measurements of the locus of contour scaling on coarse-grained brownstone faces (method described above) were made in 10 cemeteries (Figure 10.8). Idealized 'maps' of percent contour scaling averaged for a large number of coarse-grained brownstones per cemetery show inconsistent patterns from place to place, but on the most damaged stones, lower stone faces are more scaled than upper, and centres of faces more than edges. This pattern cannot be explained by mechanical abuse from cemetery maintenance (e.g. lawn mowing), which would affect lower stone corners. Lawn sprinkling, which is responsible for sandstone contour scaling in semi-arid Colorado,[72] is rare in the rainy climate of the northeastern USA.

The bottom-centre weathering pattern has been explained by capillary rise of soil moisture part way up into a porous stone, with various weathering processes occurring where evaporation of water is most intense.[73,74] In laboratory salt weathering experiments, as moisture evaporates near the upper part of its capillary potential height in stone, salts[75] or other precipitates grow and disrupt the stone. In another salt weathering experiment, Smith and McGreevy observed subflorescence in porous zones several millimetres beneath the stone surface, which caused contour scaling when the salts expanded either thermally or through hydration.[76]

Stone aspect and contour scaling

Aspect differences in sandstone tombstone weathering rates have been reported with the most affected side varying between studies.[77,78,79] Considerable aspect differences in brownstone weathering are also measured in this study. Contour scaling for all available west-facing coarse-grained stones (mean areal face loss = 32.6%, n = 281) is almost twice that for east-facing slabs (mean = 17.8%, n = 130), as previously noticed for brownstone tombstones[80] and buildings in Manhattan.[81] North-facing stones display less contour scaling (mean = 15%, n = 30). No south-facing, coarse-grained stones have been found for comparison. No consistent difference in spall thickness with aspect was observed.

DISCUSSION

This semi-quantitative geographic reconnaissance of Portland arkose 'brownstone' tombstone deterioration in the northeastern USA demonstrates that formerly high emissions of coal smoke did not influence the amount of contour scaling on the coarse-grained stones, so 'natural' causes must be invoked. Whereas brownstone exhibits a sufficient variety of texture, composition, atmospheric and ionic environment as to preclude any easy multivariate analysis of the cause of contour scaling, hypotheses about deterioration processes on an idealized stone can be formulated from field evidence and from previous literature.

A case hardened layer was evidently formed quickly after the stone was cut and polished on its flat inscribed face, and indeed on all sides of the stone. The case hardened layer is probably composed of mineral grains surrounded by secondary cement (silica?) that was removed by dissolution from the underlying feldspars and albite cement. Both quarry water and rain could have caused the dissolution and reprecipitation of silica near the evaporation surface. The sandstone underlying the case hardened layer is more porous and weaker as a result. This process occurs on many types of stone, but probably does not in itself cause the contour scaling, which requires an expansive force.

The expansive force that breaks off the case hardened layer could in part be thermally induced, derived from insolation on the dark stone (Munsell 2.5YR4/3 is typical). Although not measured, it is likely that the western face is significantly warmer than the eastern and northern faces on sunny days, especially with shading trees being relatively sparse in most cemeteries of the study region. Long-term thermal expansion and contraction alone (particularly of highly expandable quartz grains) could cause a threshold 'fatigue' point to be reached where the bonds between the case hardened layer and underlying high porosity layer, with its mica-defined bedding planes are finally broken, more on the west than on east or north sides.

Water would magnify the thermal expansive process in many ways and would be available almost constantly in the rainy eastern USA by capillary rise of soil moisture. Pure water alone can cause sandstone expansion.[82] In addition, higher evaporation rates within pores on the warmer west side should create empty pores that attract more liquid water toward the west side than toward east or north. This could cause more precipitates to form (ion source from the ground or from within the stone) on the west side. These salts, once formed would later be expanded either thermally or through increased hydration on the warmer side, aiding the shedding of the case hardened layer. Presumably, ions would precipitate in the stone at the depth where pore water evaporates due to heat penetration from the solar-heated surface. Few subflorescences or efflorescences were observed in this study, but that is because rainfall removes salts from visible surfaces.

Sandstones with fine pores have low evaporation rates.[83] That may partly explain the low contour scaling amounts on the fine-grained stones from the northern Connecticut River Valley.

A problem for any salt weathering hypothesis of brownstone tombstone weathering is to imagine a salt source in a rainy region some distance from the coast. De-icing salts would not be used in most of the cemeteries visited, but lawn fertilizers might. However, brownstones were already disintegrating in the northeastern USA before humans began any significant ion loading to rural New England. Most of the cemeteries are on recent glacial deposits (sand, till, loess), so precipitable ions are probably still available from soil weathering for inclusion in capillary water.

Certainly, shaded cooler sides (north and east) would allow inter-pore water to remain in contact with the minerals and cement a longer time, which would increase the rate of chemical weathering (dissolution, oxidation, hydrolysis), but it is likely that the total flux of water (with precipitable ions) toward an evaporation surface, rather than the duration of contact is more important for contour scaling

Driving rain, thought to have a weathering effect on the exposed stone surfaces,[84] is high in magnitude in the northeastern USA, but is at a maximum from the east,[85] which suggests it is not of major importance in brownstone decay. Likewise, the shady northern stone side, which should contain pore water from rain or capillary rise the longest, is less contour-scaled. Freezing of the moisture within the stone would also create an expansive force in the northeastern USA, but it is unclear from previous theory which side of the stone should be most affected.

If groundwater wicking is a major component of brownstone weathering, as it seems to be from 'maps' of contour scaling, then tombstones are not perfect surrogates for brownstone buildings, which display contour scaling at all heights above the ground. Whereas the tombstones are not affected by sulphur air pollution, buildings with a carbonate mortar might be. Buildings will need to be studied independently after all.

Acknowledgements

Fieldwork was partially supported by a research grant from the University of Delaware College of Arts and Science. Thanks go to two anonymous readers, who not only took the time to correct my typographical and other errors, but who forced me to think more about potential weathering mechanisms. Linda Parrish, made my hand-drafted figures electronically acceptable.

References

1. Amoroso, G.G. and Fassina, V. (1983) *Stone Decay and Conservation*. Material Science Monograph 11. Elsevier, Amsterdam.
2. Winkler, E.M. (1994) *Stone in Architecture*. Springer-Verlag, New York.
3. Julien, H. (1883) The decay of the building stones of New York City. *Proceedings American Association for the Advancement of Science* 28: 372–1980.
4. Matthias, G.F. (1967) Weathering rates of Portland arkose tombstones. *Journal of Geological Education* 15: 140–144.
5. Merrill, G.P. (1891) *Stones for Building and Decoration*. J. Wiley and Sons, New York.
6. Bell, M. (1985) *The Face of Connecticut*. State Geological and Natural History Survey of Connecticut, Bulletin 110.
7. Lewis, J.V. (1908) Building Stones of New Jersey. *Part 3, Annual Report*. Geological Survey of New Jersey: 55–124.
8. Merrill, *op.cit.* (1891)
9. Feddema, J.J. and Meierding, T.C. (1987) Marble weathering and air pollution in Philadelphia. *Atmospheric Environment* 21: 143–157.
10. Meierding, T.C. (1993) Marble tombstone weathering and air pollution in North America. *Annals, Association of American Geographers* 83: 568–588.
11. Schreiber, K. and Meierding, T.C. (1999) Spatial patterns and causes of marble tombstone weathering in western Pennsylvania. *Physical Geography* 20: 173–188.
12. Krynine, P.D. (1950) Petrology, stratigraphy and origin of the Triassic sedimentary rocks of Connecticut. *Connecticut Geological Survey Bulletin* 73. State Geological and Natural History Survey of Connecticut.
13. Lewis, *op.cit.* (1908).
14. Bell, M. (1985) *The Face of Connecticut*. State Geological and Natural History Survey of Connecticut, Bulletin 110.

15 Krynine, *op.cit.* (1950)
16 Hubert, J.F., Gilchrist, J.M., and Reed, A.A. (1982) Brownstones of the Portland Formation. In *Guidebook for Fieldtrips in Connecticut and South Central Massachusetts*. Guidebook 5. State Geological and Natural History Survey of Connecticut: 103–129.
17 Merrill, *op.cit.* (1891)
18 Krynine, *op.cit.* (1950)
19 Klein, G. (1968) Sedimentology of Triassic rocks in the lower Connecticut valley. In *Guidebook for Fieldtrips in Connecticut and South Central Massachusetts*. Guidebook 5. State Geological and Natural History Survey of Connecticut: 1–19.
20 Baedecker, P.A., Reddy, M.M., Sciammarella, C.A., and Reimann, K.J. (1990) Effects of acidic deposition on carbonate stone: Results and discussion. In Irving, P.M. (ed.) *Acidic Deposition: State of the Science and Technology*. NAPAP, Washington D.C. 19: 97–163.
21 Hubert *et al*, *op.cit.* (1982)
22 Hubert *et al*, *op.cit.* (1982)
23 Lewis, *op.cit.* (1908)
24 Merrill, *op.cit.* (1891)
25 Julien, *op.cit.* (1883)
26 Merrill, *op.cit.* (1891)
27 Matthias, *op.cit.* (1967)
28 Rahn, P.H. (1971) The weathering of tombstones and its relationship to the topography of New England. *Journal of Geological Education* 19: 112–118.
29 Geike, A. (1880) Rock-weathering as illustrated in Edinburgh church yards. *Proceedings, Royal Society, Edinburgh* 10: 518–532.
30 Winkler, E.M. (1979) Effect of case hardening in stone. *3rd International Congress on the Deterioration and Preservation of Stones*. Venice.
31 Feddema and Meierding, *op.cit.* (1987)
32 Robinson, D.A. and Williams, R.B.G. (1998) The weathering of Hastings Beds sandstone gravestones in southeast England. In Jones, M.S. and Wakefield, R.D. (eds.) *Aspects of StoneWeathering, Decay and Conservation*. Imperial College Press: 1–15.
33 Williams, R.B.G. and Robinson, D.A. (2000) Effects of aspect on weathering: anomalous behaviour of sandstone gravestones in southeast England. *Earth Surface Processes and Landforms* 25: 135–144.
34 Winkler, E.M. (1975) *Stone: Properties, Durability in Man's Environment*. Springer-Verlag, New York.
35 Winkler, *op.cit.*(1979)
36 Robinson, D.A. and Williams, R.B.G. (1987) Surface crusting of sandstones in southern England and northern France. In Gardiner, V. (ed.) *International Geomorphology. Part II*: 623–635.
37 Smith, B.J., Magee, R.W., and Whalley, W.B. (1994) Breakdown patterns of quartz sandstone in a polluted urban environment, Belfast, Northern

Ireland. In Robinson, D.A. and Williams, R.B.G. (eds.). *Rock Weathering and Landform Evolution*. J. Wiley and Sons, New York: 131–150.
38   Robinson and Williams, *op.cit.* (1987)
39   Merrill, *op.cit.* (1891)
40   Winkler, *op.cit.*(1979)
41   Matthias, *op.cit.* (1967)
42   Whalley, W.B. (1978) Scanning electron microscope examination of a laboratory-simulated silcrete. In W.B. Whalley (ed.) *Scanning Electron Microscopy in the Study of Sediments*. Geo-Abstracts, Norwich: 399–405.
43   Robinson and Williams, *op.cit.* (1987)
44   Smith *et al*, *op.cit.* (1994)
45   Williams and Robinson, *op.cit.* (2000)
46   Merrill, *op.cit.* (1891)
47   Paraguassu, A.B. (1972) Experimental silicification of sandstone. *Bulletin, Geological Society of America* 83: 2853–2858.
48   Whalley, *op.cit.* (1978)
49   Williams and Robinson, *op. cit.* (2000)
50   Winkler, *op.cit.* (1975)
51   Winkler, *op.cit.* (1975)
52   Williams and Robinson, *op.cit.* (2000)
53   Krynine, *op.cit.* (1950)
54   Geike, *op.cit.* (1880)
55   Paradise, T.R. (1999) Sandstone weathering thresholds in Petra, Jordan. *Physical Geography* 19: 205–222.
56   Merrill, *op.cit.* (1891)
57   Lehmann, E.P. (1959) The bedrock geology of the Middletown quadrangle. *Quadrangle Report Number 8*. State Geological and Natural History Survey of Connecticut.
58   Hubert *et al*, *op.cit.* (1982)
59   Yatsu, E. (1988) *The Nature of Weathering*, Sozosha, Tokyo.
60   Amoroso and Fassina, *op.cit.* (1983)
61   Hubert *et al*, *op.cit.* (1982)
62   Hosley, W. and Holcombe, S.M. (1994) By their markers ye shall know them. *The Ancient Burying Ground Association, Inc.* Hartford, CT.
63   Beers, J.B. and Company (1884) *History of Middlesex County, Connecticut, With Biographical Sketches of Its Prominent Men*. New York.
64   Julien, *op.cit.* (1883)
65   Lipfert, F.W. (1986) Estimates of historic urban air quality and precipitation acidity in selected US cities (1880–1980). *Draft Report, National Park Service Project G6–2*.
66   Schreiber and Meierding, *op.cit.* (1999)
67   Meierding, *op.cit.* (1993)
68   Feddema and Meierding, *op.cit.* (1987)
69   Lipfert, *op.cit.* (1986)
70   Merrill, *op.cit.* (1891), p. 352

71 Julien, *op.cit.* (1883)
72 Meierding, T.C. (1981) Marble tombstone weathering rates: a transect of the United States. *Physical Geography* 2: 1–18.
73 Robinson and Williams, *op.cit.* (1998)
74 Winkler, *op.cit.* (1979)
75 Goudie, A.S. (1986) Laboratory simulation of the wick effect in salt weathering of rock. *Earth Surface Processes and Landforms* 11: 275–285.
76 Smith, B.J., and McGreevy, J.P. (1988) Contour scaling of a sandstone by salt weathering under simulated hot desert conditions. *Earth Surface Processes and Landforms* 13: 697–705.
77 Geikie, *op.cit.* (1880)
78 Robinson and Williams, *op.cit.* (1998)
79 Williams and Robinson, *op.cit.* (2000)
80 Matthias, *op.cit.* (1967)
81 Merrill, *op.cit.* (1891)
82 Snethlage, R. and Wendler, E. (1997) Moisture cycles and sandstone degradation. In N.S. Baer and R. Snethlage. *Saving Our Architectural Heritage*. J. Wiley and Sons, New York: 7–24.
83 Amoroso and Fassina, *op.cit.* (1983)
84 Amoroso and Fassina, *op.cit.* (1983)
85 Sherwood, S.I. (1990) *Deposition to Structures*. National Acid Precipitation Program, State of Science and Technology Report 21. Superintendent of Documents, Washington.

11 Weathering of Portuguese Megaliths – Evidence for Rapid Adjustment to New Environmental Conditions

G.A. POPE and V.C. MIRANDA

ABSTRACT

The Alto Alentejo region of southwest Portugal is known for its collection of Megalithic monuments ($c.6500$–5500YBP). Megaliths in this region are constructed of granitic rock commonly found on an ancient weathering platform. Menhirs (vertical oblong stones, standing alone or in circular groups) appear to be no more than elongated granitic corestones unearthed, rolled into position, and uprighted. Some stones are carved, abraded, or faced, but the most common and most significant human impact is the unearthing and re-orientation of the stones. The uprighted stones often retain a soil line originating from their former half-buried position, dividing an oxidized soil side from a lichen-encrusted sub-aerial side. Recently (<10 years) unearthed field stones have a marked difference in weathering hardness on either side of the soil line. Menhirs, however, have lost this difference between former buried and subaerial sides. Instead, bimodal north-south weathering maxima are found on upright menhirs regardless of their previous orientation or burial position.

Weathering tests demonstrate that unabraded megalith surfaces, regardless of former position, are statistically indistinguishable from exposed native stone outcrops. A few thousand years of re-oriented exposure works to rewrite the face of weathering tens of millennia in the making. This brings to light the episodic nature of weathering in response to changing environments.

INTRODUCTION

As early examples of quarried stone, megaliths afford an opportunity to assess how rock adjusts to the atmospheric and biotic environment. Geomorphologists, archaeologists, and stone conservators are just now realizing the wealth of information that may be obtained by studying the deterioration of these ancient monuments.[1,2,3,4] This paper discusses weathering characteristics and relationships of granitic megaliths found on the Alta Alentejo plateau of southwest Portugal, in the vicinity of Évora and Monsaraz.

Using information on geomorphology and the weathering characteristics of the stones, one can see that Alentejan megalith stones have not been imported from long distances, and appear to be of local origin, taken from weathered granitic outcrops. Post-megalithic weathering impacts (c.6000 years) appear to override millions of years of inherited weathering accumulated on the rock prior to quarrying or unearthing. This information provides an insight into the rates and equilibrium processes of weathering.

ARCHAEOLOGICAL AND GEOMORPHIC SETTING

Archaeological setting

The European Megalithic tradition spans an era from c.7000 to 2500 YBP, from the end of the Neolithic into the Calcolithic periods.[5,6,7] Megalithic monuments are therefore not diagnostic of a specific culture,[8] although they are associated with the rise of organized agriculture in Western Europe.[9] Since Iberia was a seat of initial agriculture development in Europe, it has been suggested that the entire megalithic tradition emanated from Iberia as well.[10,11] Aside from some engraving and a few examples of shaping, most of the Alentejo megaliths are barely altered, unlike the roughly rectangular hewn stones known at Stonehenge and Malta or the graceful ellipsoids of Carnac. Carvings on some megaliths are similar across the region, irrespective of chronology.[12] Megalith development is said to have evolved during several phases.[13] Cromeleque dos Almendres (near Évora) had three separate phases of construction from Early to Late Neolithic,[14] and re-facing of some menhirs at Vale Maria de Meio (also near Évora) occurred late in the Neolithic.[15] This puts the age of these monuments at roughly 5500–6500YBP. Further details on the megalithic archaeology of the Évora region may be found in Sarantopoulos.[16]

Portuguese megaliths may be classified by type into (1) single oblong or rectangular standing stones (menhirs), (2) closed circles, ellipses, or squares of smaller menhirs (cromeleques in Portuguese, Figure 11.1), and (3) chambers constructed of large leaning rock slabs (dolmens, known as antas in Portuguese, Figures 11.2 and 11.3). Archaeologists agree that megaliths probably had ceremonial significance. Antas are known to be burial chambers; cromeleques are oriented in cardinal directions and may have been used in astronomical calculations; single menhirs may have been used for surveys or landmarks.[17,18,19] Menhirs and cromeleques are free-standing (or toppled over the years). Many antas, however, were buried for some period. Exposed antas today were either excavated or their cover mounds eroded over time.

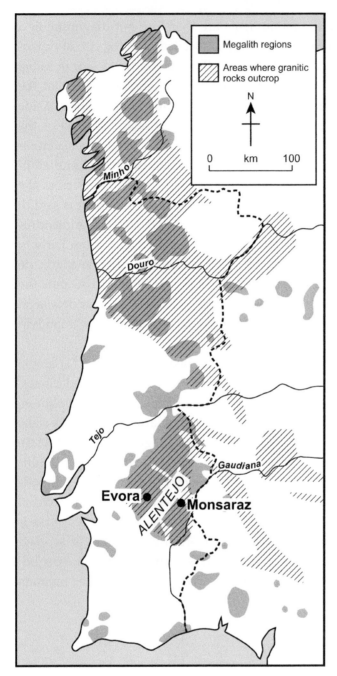

Figure 11.1 Locations of megalithic monuments in western Iberia. Locations where granitic rock may be found are also shown, and tend to correspond to megalithic areas.

WEATHERING OF PORTUGUESE MEGALITHS 203

Figure 11.2 Cromeleque and menhir at Xerez, near Monsaraz. Centre menhir is approximately 4 m high.

Figure 11.3 Partially toppled anta at Pinheiro do Campo, near Évora. Front slab is approximately 2 m high.

Geomorphic setting

The western Iberian Peninsula is dominated by folded Palaeozoic metamorphic rocks with regions of plutonic rock emplaced during the Hercynian (also known as Variscan, late Palaeozoic) orogeny.[20] Regional maps portray late-tectonic granitic masses dispersed through Carboniferous metasediments across Alto Alentejo.[21] Some folding and faulting occurred with the later Alpine orogeny, but the region remained relatively undeformed. As a result, the bedrock was able to weather and erode during the Mesozoic and Cenozoic eras.

There are few geomorphic studies of the Alentejo region published outside of Portugal.[22,23,24] In his regional geography monograph, Houston[25] cites Feio[26] in explaining the Alentejo tableland as a Tertiary erosion surface, dissected at the edges of the plateau and near the major river valleys. Long-term weathering of granites is recognized across Atlantic Europe, including coastal and northern Portugal[27] and western Spain[28] to produce deep saprolites ('arenes'). The granite weathering profiles identified in the Évora and Monsaraz regions currently fall under a more xeric climate. If we accept Sequeira Braga's et al. hypothesis that mid-latitude weathering profiles are controlled by temperature,[29] then the Alentejo saprolites probably most resemble those of western Spain.[30] There, the weathering mantle is said to be up to 58 million years old, with secondary weathering mantles produced during the middle and late Tertiary periods. More resistant granite produces spheroidally weathered corestones and, in larger masses, inselbergs and tors.[31] These are commonly found in the granitic landscapes of Alentejo. Exactly how long these remnant rock masses have been exposed is unknown.

The Alentejo region presently exists in a 'Mediterranean-Iberoatlantic' climatic province with subhumid, winter-dominant precipitation (600–900 mm).[32] In the Neolithic era of megalith building in Iberia, Portugal was in the midst of the Climatic Optimum (Atlantic Period).[33] The Alentejo region (and most of Iberia) contained mixed deciduous forest, indicative of warm but perhaps wetter conditions.[34] Since that time, Iberian climates probably became drier, then cooler and wetter, before ameliorating to present conditions. In any event, the past 6000 years did not see the extreme changes of the earlier Holocene and Late Pleistocene, nor were the changes as dramatic as those in Northwest Europe.[35]

METHODS

Twelve sites were surveyed in two regions of Alto Alentejo: Évora, and Monsaraz (Table 11.1). Non-megalith field stones at four sites, similar in shape and size to the stones used in megaliths, provided a baseline for comparison with the megaliths. Megalith rock type was recorded in the field, and compared to nearby natural rock outcrops. Further laboratory petrographic characterization was not attempted, as it was not possible to collect samples from the megaliths because of their archaeological sensitivity.

The type and degree of weathering provided evidence of exposure history and human alteration of megalithic stones. Categorical information was recorded at each stone: presence and extent of lichens, discolouration (e.g. from oxidation), weathering morphology (such as spalls, pits, fissures, and granular disintegration), and obvious human alteration such as carving, abrasion, or dressing. Orientation with respect to solar radiation was recorded. While most of the megalith sites surveyed were in a state of deterioration, several had been excavated and partially restored during recent archaeological surveys. This presented a problem for establishing the original placement and orientation of the stones (particularly the larger menhirs). Stones that were obviously displaced from their original position were not included as part of the orientation data subset. All other stones were assumed to be in their original position, or if restored, placed in their correct orientation by archaeologists.

Quantitative weathering data were derived with a type-L Schmidt hammer. The Schmidt hammer recorded the rebound of a constant

Table 11.1 Locations surveyed for this study.

Feature	Location	ID	Number of stones*
Cromeleque dos Almendres	Guadalupe, Évora	EV-CDA	5
Anta do Zambujeiro	Valverde, Évora	EV-ADZ	4
Cromeleque de Portela de Mogos (incl. 1 field stone)	São Matias, Évora	EV-PDM	4
Menhir do Oliveirinha	Graça do Divor, Évora	EV-OLI	1
Cromeleque Vale Maria do Meio	São Matias, Évora	EV-VMM	2
Anta de Pinheiro do Campo (incl. 1 field stone)	Giesteira, Évora	EV-APC	3
Menhir do Casbarra	São Matias, Évora	EV-CAS	1
Menhir do Outiero	Outiero, Monsaraz	MS-OUT	1
'Menhir' at Sao Brisos†	São Brisos, Monsaraz	MS-SBR	1
Cromeleque de Xerez,	Xerez, Monsaraz	MS-XER	2
Anta 2 do Olival de Pega	Telheiro, Monsaraz	MS-AOP	2
Field stones (no megaliths)	São Pedro do Corval, Monsaraz	MS-SPC	4
Quarried granite	Valverde, Évora	EV-VGQ	1

Menhir = single oblong standing stone; generally >2 m in height, may be toppled
Cromeleque = elliptical, circular, or rectangular alignment of standing stones (menhirs), most stones <2 m in height, but may be toppled
Anta = dolmen tomb, slabs of rock set on edge, capped with 1 or more roof slabs
Field stones = naturally occurring, exposed remnants from bedrock; not megaliths
* 'Number of stones' refers to the number surveyed, some locations had many more stones
† The 'menhir' at São Brisos is actually a 'mushroom rock' excavated to a depth of ~3 m. It is essentially an exposed granite bedrock outcrop, and not at all similar to other oblong menhirs.

force impact on the rock, thereby measuring rock structural integrity. Weathering reduces the structural integrity by decomposition or increases structural integrity if precipitated weathering products indurate the rock surface. Examples of Schmidt hammer use in geomorphic contexts are presented in Day, Sjöburg , and Tang .[36,37,38] Schmidt hammer testing is problematic for coarse-grained rocks (such as granites), in that the large crystals of varying mineralogy tend to yield data with a broad statistical spread. At least ten readings, sampled within a confined area on the rock (c.50'50 cm) were necessary to account for statistical spread. Other methods of weathering assessment achieve better data on megalith weathering (cf. Delgado Rodrigues for tomography,[39] Silva et al. for petrographic analysis,[40] and Sellier for surface recession[41]). However, the type-L Schmidt hammer is considered a 'low impact' testing device, resulting in little or no visible scarring on soft or sensitive materials. Furthermore, the Schmidt hammer has advantages over other assessment methods in that it is inexpensive, portable, provides consistent quantitative data, and does not require elaborate set-up or physical samples.

RESULTS AND DISCUSSION

Source material

Weathered granitic rock forms a major portion of the landscapes surrounding the megaliths of Évora and Monsaraz. It is interesting to note that megalith concentrations throughout western Iberia tend to appear in areas where granitic rocks outcrop nearby (Figure 11.1). We cannot speculate on whether granite was somehow significant to megalith builders, but we did note that granitic rock seems to be the stone of preference in the Alto Alentejo, despite the availability of other rock types. Western European megaliths are composed of many rock types, although granitic landscapes are prominent in two noteworthy megalith regions, Brittany[42] and Dartmoor.[43]

While ancient granitic quarry sites for the region are uncertain,[44] natural outcrops are abundant. Ample material for megalith construction existed with the remnant corestones and exfoliation slabs eroding out of the regolith and bedrock. Barely altered and encrusted with lichens, an untrained eye could mistake a collapsed dolmen for an angular tor, or a fallen menhir as one of the ubiquitous corestones. Kalb stated that specific types of stone were imported over distances up to 8 km for megaliths at Vale de Rodrigo (southwest of the Almendres and Zambujeiro sites).[45] This was consistent with our observations, particularly at Cromeleque dos Almendres, where several types of granite were noted. Criado Boado and Fábregas Valcarce contend that adjacent outcrops were seldom used as quarries for megaliths in Galicia.[46] However, minimal damage to surface weathering features on most megalith stones (discussed below) suggested that the stones were not dragged or rolled long distances (at least not without the aid of sledges or other mechanical aids).

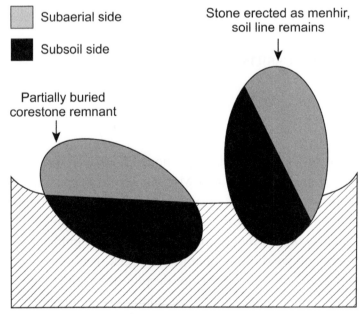

Figure 11.4 Schematic of remnant corestones and their configuration as menhirs.

Therefore, we feel that most source material (or all source material on smaller monuments) was derived from nearby outcrops if not on-site (particularly in the Monsaraz region).

One of the most visible features on the individual menhir and cromeleque stones was the 'soil line', which vertically divided the former subaerial (above soil) side from the buried side of the upended stones (Figure 11.4 schematic, Figure 11.5 photograph). We estimate that these buried sides have been exposed to air for over 5000 years, based on the supposed age of the monuments.[47,48,49] On stones not polished, dressed, or engraved, the former subaerial side often retains a biotic growth of lichen, while the former buried side exhibits an oxidized orange colour. The soil line between these disparate zones is usually prominent, and on only a few stones are formerly buried areas colonized by lichen. Given this dichotomous appearance, Schmidt hammer tests were used to determine if

Figure 11.5 Small (1.5 m) menhir at Cromeleque dos Almendres, with a planed off top and polished "cup" depressions. The planed top and cups were distinctly harder than the rest of the rock. This stone also shows a distinct lichen-encrusted former subaerial side (to the right) and a pale, oxidized buried side (to the left). The flat-faced menhir in the background has also been planed off.

post-Megalithic weathering is lithologically or environmentally controlled.

Weathering of megaliths

Superficial weathering features on the megaliths were similar to those seen on field stones. While no attempt was made to specifically differentiate between weathering processes, it was apparent that several weathering processes were active. Oxidation of biotite and dissolution of feldspars and biotite contributed to granular disintegration. Several megaliths were extensively weathered by granular disintegration, notably at São Brisos and at Cromeleque dos Almendres. Weathering pits were less common on fieldstones, cromeleques, and single menhirs, but were more commonly observed on the slabs used in antas. Pressure unloading dilation combined with the weakening effects of chemical weathering to produce exfoliation and spalling (2–20 mm thick, based on scars on the rocks, but of unknown areal dimension). Natural joints provided avenues for further weathering, and often exhibited oxidation. Freeze-thaw mechanisms could have contributed to mechanical weathering during occasional winter cold spells. In the relatively sunny climate, solar heating probably acted as a catalyst for chemical weathering if not to thermal expansion mechanical weathering. Finally, anthropogenic impacts such as abrasion and carving altered the weathering environment. Polishing on some stones created a protective silica seal on the granite, and stones that appeared polished were resistant to weathering. Many of the superficial weathering forms observed here were also observed by Sellier on granitic megaliths at Carnac, though not to the advanced degree observed in that region.[50] The differences may be due to rock mineral composition and/or the wetter climate of northwest France. Sellier did not make any comparisons with natural field stones.[51]

Before discussing correlations between weathering and various weathering factors, it is necessary to mention relationships between these parameters, and to test for covariance. These comparisons are represented by chi-square cross-tabulations in Table 11.2. Random testing insured that data from megaliths and field stones were not preferentially oriented to an exposure direction, nor was there a preference in sampling former subaerial or former buried surfaces. Though not exclusively, engravings and stone dressing were marginally related to exposure direction ($c^2 = 16.561$, $p = 0.172$). Engravings and facing tended to face southeast to southwest; and this may be relevant in that these stones could have been placed for astronomical observations.[52] Engravings and stone dressing were not preferentially placed on buried or subaerial surfaces ($c^2 = 0.037$, $p = 0.694$).

Almost exclusively, lichens were confined to former subaerial surfaces ($c^2 = 27.623$, $p = 0$). As expected, carved and dressed surfaces had very little lichen growth ($c^2 = 20.703$, $p = 0$), and on both field stones and unaltered megalith stones, lichen occurrence was essentially similar ($c^2 = 1.874$ $p = 0.392$). Comparisons between lichen cover and exposure orientation were statistically significant; most lichen-bare faces on unaltered stones were preferentially oriented to the south and west ($c^2 = 30.546$, $p = 0.015$). This lichen orientation may reflect a purposeful placement, but more likely the relationship reveals an ecological preference of lichen growth toward cooler, moister conditions to the north and northeast.

While weathering can produce hard, indurated surfaces, R-value data were consistent with a reduction in structural integrity due to weathering. Individual R-values (each individual hammer test) and mean R-values (means of 10 or more hammer tests on one area of the stone) were compared against weathering factors such as degree of lichen cover, presence of human alteration (dressing, polishing, or engraving), type of stone (megalith vs. field stone), former soil position (buried vs. exposed subaerial), and exposure orientation (after construction). These comparisons are summarized in Table 11.3. Mean R-values tended to be less than statistically significant, while the set of individual (not averaged) R-values did reveal statistical significance with several weathering factors. Ordinarily, we favour mean data over individual data because mean data compensate for

Table 11.2 Chi-square cross-tabulation between weathering factors, testing for independence between factors. Each comparison of factors lists number of cases (N), Pearson's chi-square value (χ^2), and probability of error (p). Factors that are statistically independent are in boldface.

	Soil position (subaerial/buried)	Human alteration (altered/unaltered)	Lichen coverage (1 least – 3 most)	Cardinal exposure orientation
Type of stone (megalith/field stone)	N = 44 $c^2 = 0.155$ $p = 0.694$	N = 66 no field stones were altered	N = 63 $c^2 = 1.874$ $p = 0.392$	N = 45 $c^2 = 2.485$ $p = 0.962$
Soil position (subaerial/buried)		N = 44 $c^2 = 0.037$ $p = 0.694$	N = 44 $c^2 = 27.623$ $p = 0.000$	N = 30 $c^2 = 13.225$ **$p = 0.040$**
Human alteration (altered/unaltered)			N = 63 $c^2 = 20.703$ $p = 0.000$	N = 45 $c^2 = 16.561$ $p = 0.172$
Lichen coverage (1 least – 3 most)				N = 45 $c^2 = 30.546$ **$p = 0.015$**

WEATHERING OF PORTUGUESE MEGALITHS 213

Table 11.3 Analysis of variance (ANOVA) statistical comparison of rock hardness (R-value) against weathering factors, stating number of cases (N), ANOVA F statistic, and probability of error (p). 'Mean R-values' are calculated with a set (usually 10) of hammer tests per rock location. 'Individual R-values' consider each hammer strike individually. Statistically significant relationships ($p<0.100$) are shown in boldface.

Data set	Type of stone (megalith/ field stone)	Soil position (subaerial/ buried)	Human alteration (altered/unaltered)	Degree of lichen coverage (1–3)	Cardinal exposure orientation
Full data set mean R-values (from each stone)	N =66 F = 0.301 p = 0.585	N = 44 F = 0.013 p = 0.960	N = 66 F = 0.690 p = 0.409	N = 63 F = 0.914 p = 0.407	N = 45 **F = 1.862** ***p* = 0.097**
Full data set, individual R-values	N = 692 F = 2.032 p = 0.155	N = 451 F = 0.941 p = 0.332	N = 692 **F = 5.255** ***p* = 0.022**	N = 656 **F = 10.271** ***p* = 0.000**	**N = 419*** **F = 10.158** ***p* = 0.000**
Unaltered stones mean R-values (from each stone)	N = 45 F = 0.527 p = 0.472	N = 40 F = 0.056 p = 0.815	—	N = 45 F = 0.091 p = 0.913	N = 27 F = 1.197 p = 0.340
Unaltered stones individual R-values	N = 487 F = 1.117 p = 0.291	N = 427 F = 1.083 p = 0.299	—	N = 487 F = 2.248 p = 0.107	N = 302 **F = 8.612** ***p* = 0.000**
Unaltered megaliths only mean R-values (from each stone)	—	N = 28 F = 0.217 p = 0.645	—	N = 32 F = 0.020 p = 0.981	N = 25 F = 0.866 p = 0.501
Unaltered megaliths only, individual R-values	—	N = 325 F = 0.043 p = 0.835	—	N = 550 F = 3.341 p = 0.068	N = 386 **F = 9.569** ***p* = 0.000**
Field stones only mean R-values	—	N = 9 F = 0.143 p = 0.716	—	N = 10 F = 0.473 p = 0.641	(insufficient data)
Field stones only individual R-values	—	N = 104 **F = 6.189** ***p* = 0.014**	—	N = 116 **F = 7.530** ***p* = 0.001**	N = 23 **F = 4.234** ***p* = 0.052**

the variation in R-value readings caused by the heterogeneity of granitic rock. However, the statistical relationships revealed by the individual R-value data set are intriguing, and warrant further discussion.

Biotic growths (such as bacteria, algae, lichens) are known to contribute to weathering, with penetrating root hyphae and release of powerful chemical weathering agents such as chelates and organic acids;[53] Romão and Rattazzi discussed the rapid biodeterioration on megalithic tombs near Évora.[54] As expected, we found a decrease in structural integrity (in the individual R-value data set) in areas with more lichen. Differences in R-values were greatest between lichen class 2 and 3 (some to extensive lichens) and lichen class 1 (no lichens).

Megalithic stones found throughout Europe often exhibit engravings and decoration.[55] Of the megalithic stones surveyed here, most lacked visible human alteration. This may be due to weathering, which obliterated carved surface features, or may be due to the fact that not all stones were engraved or altered. Data indicated that weathering did not completely erase such carvings over the period of exposure (c.6000 years). Engravings seen at the Almendres, Portela de Mogos, and Xerez cromeleques and Casbarra and Outiero single menhirs were muted in appearance, but still visible. A few stones exhibited apparent stone dressing or abrasion, with flat, sometimes polished, surfaces (Figure 11.5). There was some difference in R-values between altered vs. unaltered stones. The difference was statistically significant for the entire data set of individual R-values. As expected, carving and stone dressing removed weathered surface material, exposing harder material. Pope presented similar findings regarding the weathering of petroglyphs.[56]

There was no statistically significant difference in R-values, individual or mean, between megalith stones and field stones, implying that weathering impact was similar between the two. Two explanations could account for this correspondence:

1. weathering processes were slow, and weathering since construction was insufficient to override the inherited impacts of millions of years of *in situ* weathering; or
2. weathering was relatively rapid and the megalith stones quickly reached an equilibrium weathering state similar to the native field stones.

We favour the second scenario, based on soil position and exposure data, discussed below.

R-values were statistically similar between surfaces formerly buried and surfaces that were exposed to the air prior to megalith construction (Figure 11.6). Formerly buried anta slabs were equally as weathered as formerly subaerial cromeleque and menhir stone

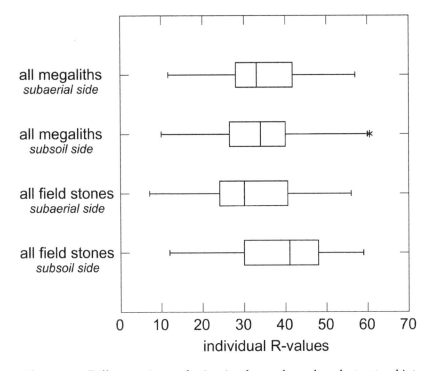

Figure 11.6 Differences in weathering (as shown through rock structural integrity R-values) between subaerial and buried sized of the unaltered megaliths and field stones. There is no statistical difference between subaerial and buried sides on megaliths. The difference is statistically significant on field stones.

surfaces. The unaltered field stones (not incorporated into megaliths) still exhibit marked differences in weathering between subaerial and buried sides (ANOVA $p = 0.014$), with more weathering on the subaerial side. On native field stones, subaerial weathering is apparently stronger than weathering in the soil environment. The important difference to note here is that field stones have not been exposed to the extremes of subaerial weathering experienced by all sides of the megalith stones for several millennia after emplacement.

Degree of weathering varies according to exposure orientation (after emplacement). Exposure orientation is a significant factor, overriding the former soil position factor. Multiple variables come into play with the orientation factor: differences in lithology from one side of the stone to the other; shade; direction of precipitation; and solar angle. R-values on different exposures were generally lower in the quadrant from southwest to northwest (Figure 11.7). Three single stone exceptions with unique exposures (the centre menhir at Cromeleque de Portela de Mogos, and two peripheral menhirs and Cromeleque dos Almendres) have anomalously high and variable R-values, departing from the trend. Excluding these anomalous stones, exposure orientation is statistically significant as a factor for

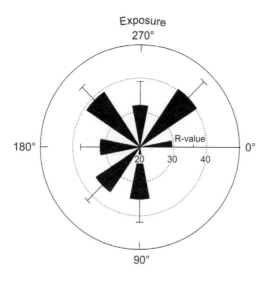

Figure 11.7 Differences in weathering (as shown through rock structural integrity R-values) at different exposures for a subset of megalith stones (n = 364). Lower R-values (lower structural integrity, hence, more weathered) appear in the quadrant from southwest to northwest.

R-value for every subset of individual R-values. Many authors note a weathering preference to the south or southwest in the mid- to low-latitudes of the northern hemisphere.[57,58,59] The weathering maxima we observe from the southwest to north may be due to a combination of afternoon insolation and direction of precipitation from passing weather systems[60] or propensity for frost on more northern exposures.[61]

The period of exposure for the megaliths included times of wetter and/or cooler climates in the Iberian region.[62] At present, we cannot test for the degree of influence of these potentially weathering-enhancing climatic periods, or the translation to microclimates most important to weathering,[63] though it would be an interesting investigation. It is reasonable, however, to conclude that the weathering we see today on the megaliths reflects the combination of environments over the past 6500 years and the initial characteristics of the corestones from which they derive, which in turn represent perhaps millions of years of environmental change. Our data presented here indicate the significance of more recent environments over the inherited effects from older environments.

CONCLUSIONS

The granitic megalith stones of the Alto Alentejo are interesting because of what they reveal about weathering rates. From this information, we can speculate about their origin and construction, and recommend practices for their conservation. The results are summarized below:

♦   Naturally weathered outcrops provided material for these early megalith monuments, a practice possibly used in megalith construction across western Europe. Lack of damage to superficial weathering features suggests that, despite evidence of importation into locations of differing lithology,

the megalith stones were not transported long distances, or alternatively, were transported with great care.
- Weathering processes active on the megaliths included biotic, mechanical, and microclimate-influenced dissolution processes. While there is evidence of some human alteration (in the form of engravings or dressing) that removed original surfaces, most megaliths in the Alentejo region have superficial weathering features similar to the local field stones. Schmidt hammer data on rock hardness corroborate these results. Except where altered, megalith stones are statistically identical in weathering-controlled rock hardness to natural field stones. Stone dressing and polishing remain relatively clear, and engravings, while muted, are visible after more than 5000 years of exposure.
- Visually, megalith stones still retain former subaerial and buried sides, despite their current placement. Lichens grow on surfaces formerly situated on the subaerial side of the stone, while oxidation staining prevails on the former buried side of the stone. As expected, lichens promote weathering and contribute toward a reduction in structural integrity. Lichen colonization is an obvious concern for conservators, but eradication can be a problem if doing so damages the stone surface and any engravings. Where lichens are not present, there is no difference in weathering (as detected through rock structural integrity) between former subaerial and former buried sides, counter to what might be expected (and what appears on recently unearthed non-megalith field stones).
- Post-emplacement exposure may be a factor in the degree of weathering. There is a preference for a reduction in structural integrity in a quadrant from southwest to northwest, independent of the presence of lichens or former subaerial or buried characteristics. This exposure factor cannot have existed prior to megalith construction, and suggests that post-megalith weathering overrides characteristics inherited over a much longer pre-megalith weathering interval (and over varying

climates and microclimates). Conservators can anticipate areas of concern on certain exposures, particularly after ruined monuments have been excavated and reconstructed.

As the first public monuments of humankind,[64] megaliths provide unique opportunities to extend geomorphic theory and conservation practice. Both geomorphic and built, megaliths exist at an age that promotes translation between studies of more recent building stone and more ancient natural landscapes. Further investigations in different climates (e.g. Brittany, Cornwall, Ireland, or Malta) and with different types of stone (sandstone, slate, etc.) can expand on the results presented in this initial investigation.

Acknowledgements

We would like to thank J. Delgado Rodrigues at LINC (Portugal) for advice and direction in the field, and J. Zilhão, IPAR (Portugal), for guidance over several years of work in Portugal. Thanks also to R. Rubenstein and S. McBride for assistance with data collection. K. Irvine, A. Turkington, and several anonymous reviewers provided helpful comments. Fieldwork for this project was supported by grants from the Global Education Centre and Office of Sponsored Projects at Montclair State University.

References

1. Delgado Rodrigues, J. (1994) Conservation of granitic rocks with application to the megalithic monuments. In Delgado Rodrigues, J. (ed.) *Degradation and Conservation of Granitic Rocks and Monuments.* European Commission, Brussels: 161–246.
2. Silva, B.M., Rivas, T., Prieto, B., Bello, J.M. and Carrera, F. (1994) The Dombate Dolmen: A case study on the development of conservation techniques for megalithic monuments. In Delgado Rodrigues, J. (ed.) *Degradation and Conservation of Granitic Rocks and Monuments.* European Commission, Brussels: 367–372.

3   Romão, M.S.P. and Ratazzi, A. (1995) Biodeterioration on megalithic monuments: Study of lichen's colonization on Tapadão and Zambujeiro dolmens (Southern Portugal). International *Biodeterioration and Biodegradation* 37: 23–35.
4   Sellier, D. (1997) Utilisation de mégalithes comme marqueurs de la vitesse de l'erosion des granites en milieu tempéré: enseugnements apportés par les Alignements de Carnac (Morbihan). *Zeitschrift für Geomorphologie*, N.F. 41: 319–356.
5   Daniel, G. (1958) The Megalith Builders of Western Europe. Hutchinson, London.
6   Service, A. and Bradbery, J. (1979) *Megaliths and Their Mysteries*. Macmillan Publishing Company, New York.
7   Joussaume, R. (1988) *Dolmens for the Dead*. [translation by A. Chippindale and C. Chippindale]. Cornell University Press, Ithaca.
8   Sherratt, A. (1995) Instruments of conversion? The role of megaliths in the Mesolithic/Neolithic transition in North-West Europe. *Oxford Journal of Archaeology* 14: 245–260.
9   Joussaume, *op. cit.* (1988)
10 MacKie, E. (1977) *The Megalith Builders*. Oxford University Press, Oxford
11 Joussaume, *op. cit.* (1988)
12 Bueno Ramirez, P. (1992) Le plaques décorées Alentéjaines: Approache de leur étude et analyse. *L'Anthropologie* (Paris) 96: 573–604.
13 Holtorf, C.J. (1998) The life-histories of megaliths in Mecklenburg-Vorpommern (Germany). *World Archaeology* 30: 23–38.
14 Gomes, M.V. (1997) The Cromlech of Almendres: One of the first public monuments of mankind. In Sarantopoulos, P. (ed.) *Paisagens Arqueologicas a Oestede Évora*. Câmara Municipal de Évora, Évora, Portugal: 33–34.
15 Gomes, M.V. (1997) The Cromlech of Portela de Mogos: A socio-religious monument. In Sarantopoulos, P. (ed.) *Paisagens Arqueologicas a Oestede Évora*. Câmara Municipal de Évora, Évora, Portugal: 39.
16 Sarantopoulos, P. (1997) *Paisagens Arqueologicas a Oestede Évora*. Câmara Municipal de Évora, Évora, Portugal.
17 Daniel, *op. cit.* (1958)
18 Service and Bradbery, *op. cit.* (1979)
19 Joussaume, *op. cit.* (1988)
20 Krebs, W. (1976) The tectonic evolution of Variscan Meso-Europa. In Ager, D.V. and Brooks, M. (eds.) *Europe from Crust to Core*. John Wiley and Sons, Chichester: 119–139.
21 Congrés Géologique International (1981) *Carte Tectonique Internationale de l'Europe et des Régions Avoisinantes, Feuille 13* (1:2,500,000). USSR Academy of Sciences (Directorate of Geodesy and Cartography) and UNESCO, Moscow.
22 Feio, M. (1952) *A Evolucão do Relevo do Baixo Alentejo d Algarve; Estudo de Geomorphologia*. Centro Estudio Geographia, Lisbon.

23   Martins, A and Barbosa, B.P. (1992) Planaltos do nordeste da Bacia Terciaria do Tejo (Portugal). *Communicacoes dos Sevicos Geologicos de Portugal* 78: 13–22.
24   Pimentel, N. and Azevedo, T.M. (1992) Sobre e evolucão do relevo e sedimentacão Quaternaria do Baixo Alentejo (Alvalade-Panoias). *Gaia* (Lisbon) 4: 21–24.
25   Houston, J.M. (1967) *The Western Mediterranean World*. Frederick A. Praeger, New York. pp. 175–176
26   Feio, *op. cit.* (1952)
27   Sequeira Braga, M.A., Lopes Nunes, J.E., Paquet, H. and Millot, G. (1990) Climatic zonality of coarse granitic saprolites ('arenes') in Atlantic Europe from Scandinavia to Portugal. In V.C. Farmer and Y. Tardy (eds.) Proceedings of the 9th International Clay Conference, Strasbourg, 1989, vol. I, Clay-Organic Interactions, Clay Minerals in Soils. *Sciences Géologiques, Mémoire* 85: 99–108.
28   Molina, E., Blanco, J.A., Pellitero, E. and Cantano, M, (1987) Weathering processes and morphological evolution of the Spanish Hercynian massif. In Gardner, V. (ed.) *International Geomorphology (part II)*. John Wiley and Sons, Chichester: 957–977.
29   Sequeira Braga's *et al, op. cit.* (1990)
30   Molina *et al, op. cit.* (1987)
31   Twidale, C.R. (1982) *Granite Landforms*. Elsevier, Amsterdam.
32   Gomes, C.J.P. (1997) Ecological outline and phytosociological considerations. In Sarantopoulos, P. (ed.) *Paisagens Arqueologicas a Oestede Évora*. Câmara Municipal de Évora, Évora, Portugal: 7–13.
33   Bell, M. and Walker, M.J.C. (1992) *Late Quaternary Environmental Change: Physical and Human Perspectives*. John Wiley and Sons, New York.
34   Huntley, B and Birks, H.J.B. (1983) *An Atlas of Past and Present Pollen Maps of Europe 0–13,000 Years Ago*. Cambridge University Press, London.
35   Bell and Walker, *op. cit.* (1992)
36   Day, M. J. (1980) Rock hardness: field assessment and geomorphic importance. *Professional Geographer* 32: 72–81.
37   Sjöburg, R. (1994) Diagnosis of weathering on rock carving surfaces in Northern Bohuslän, Southwest Sweden. In Robinson, D.A. and Williams, R.B.G. (eds.) Rock Weathering and Landform Evolution. John Wiley and Sons, Chichester: 223–41.
38   Tang, T. (1998) Field testing of rock hardness and its relationship to limestone dissolution in Guilin, Southern China. *Middle States Geographer* 31:15–22.
39   Delgado Rodrigues, *op. cit.* (1994)
40   Silva *et al, op. cit.* (1994)
41   Sellier, *op. cit.* (1997)
42   Sellier, *op. cit.* (1997)
43   Bradley, R. (1998) Ruined buildings, ruined stones: Enclosures, tombs, and natural places in the Neolithic of South-West England. *World Archaeology* 30: 13–22.

44 J. Delgado Rodrigues, pers. comm.
45 Kalb, P. (1996) Megalith-building, stone transport and territorial markers: Evidence from Vale de Rodrigo, Évora, South Portugal. *Antiquity* 70: 683–685.
46 Criado Boado, F. and Fábregas Valcarce, R. (1994) Regional patterning among the megaliths of Galicia (NW Spain). *Oxford Journal of Archaeology* 13: 33–47.
47 Calado, M. (1997) Vale Maria do Meio and Neolithic cultural landscapes. In Sarantopoulos, P. (ed.) *Paisagens Arqueologicas a Oestede Évora*. Câmara Municipal de Évora, Évora, Portugal: 49–51.
48 Gomes, M.V. (1997) The Cromlech of Almendres: One of the first public monuments of mankind. In Sarantopoulos, P. (ed.) *Paisagens Arqueologicas a Oestede Évora*. Câmara Municipal de Évora, Évora, Portugal: 33–34.
49 Gomes, M.V. (1997) The Cromlech of Portela de Mogos: A socio-religious monument. In Sarantopoulos, P. (ed.) *Paisagens Arqueologicas a Oestede Évora*. Câmara Municipal de Évora, Évora, Portugal: 39.
50 Sellier, *op. cit.* (1997)
51 Sellier, *op. cit.* (1997)
52 Service and Bradbery, *op. cit.* (1979)
53 Wakefield, R.D. and Jones, M.S. (1998) An introduction to stone colonizing micro-organisms and biodeterioration of building stone. *Quarterly Journal of Engineering Geology* 31: 301–313.
54 Romão and Rattazzi, *op. cit.* (1995)
55 Bueno Ramirez, *op. cit.* (1992)
56 Pope, G.A. (2000) Weathering of petroglyphs: direct assessment and implications for dating methods. *Antiquity* 74: 833–843.
57 Paradise, T.R. (1995) Sandstone weathering thresholds in Petra, Jordan. *Physical Geography* 16: 205–222.
58 Warke, P.A., Smith, B.J. and Magee, R.W. (1996) Thermal response characteristics of stone: Implications for weathering of soiled surfaces in urban environments. *Earth Surface Processes and Landforms* 21:295–306.
59 Robinson, D.A. and Williams, R.B.G. (1999) The weathering of Hastings Beds sandstone gravestones in South East England. In Jones, M.S. and Wakefield, R.D. (eds.) *Aspects of Stone Weathering, Decay, and Conservation. Proceedings of the 1997 Stone Weathering and Atmospheric Pollution Network Conference (SWAPNet '97)*. Imperial College Press, London: 1–15.
60 Paradise, *op. cit.* (1995)
61 Meierding, T.C. (1993) Inscription legibility method for estimating rock weathering rates. *Geomorphology* 6: 273–286.
62 Bell and Walker, *op. cit.* (1992)
63 Pope, G.A., Dorn. R.I. and Dixon, J.C. (1995) A new conceptual model for understanding geographical variations in weathering. *Annals of the Association of American Geographers* 85: 38–64.

64 Gomes, M.V. (1997) The Cromlech of Almendres: One of the first public monuments of mankind. In Sarantopoulos, P. (ed.) *Paisagens Arqueologicas a Oestede Évora*. Câmara Municipal de Évora, Évora, Portugal: 33–34.

[Previously published in modified form as 'A Geomorphology of Megaliths: Neolithic Landscapes in the Alto Alentejo, Portugal', by Gregory A. Pope and Vera C. Miranda, *Middle States Geographer*, 1999, v. 32: 110–124]

12 Influence of Anthropogenic Factors on Weathering of the Carpathian Flysch Sandstones

W. WILCZYŃSKA-MICHALIK and
M. MICHALIK

ABSTRACT

The rate of weathering in a polluted atmosphere increases significantly in comparison with natural weathering. The higher weathering rate is related to the concentration of aggressive gaseous components of air pollution and the presence of anthropogenic dust particles and the composition of rocks and their texture.

Different parts of the Beskidy Mountains (the Outer Carpathians) have been chosen for sampling due to differences in composition and concentration of dust and gases in the atmosphere. Samples were collected from surface layers of sandstone in natural outcrops and buildings in which sandstone of local origin had been used as a building material. The presence of atmospheric pollution is marked on the rock surface by occurrences of gypsum crusts and efflorescences, and anthropogenic dusts.

Only locally, in the western part of Carpathians situated close to industrial areas, does salt weathering related to a high concentration of atmospheric pollution significantly accelerate the decay of rocks.

Removal of rock components during blistering and exfoliation of gypsum crusts and granular disintegration related to crystallization of numerous disseminated gypsum crystals are the most important processes of rock decay. Where the concentration of air pollution is lowest only dispersed gypsum crystals are present on the rock surface.

Rainwater of relatively low pH and low concentration of dissolved components inhibits the formation of gypsum crusts and facilitates the dissolution of primary cements in the rock. The absence of gypsum is typical of areas of very low air pollution. However, newly formed minerals of low solubility, such as barite, testify to the presence of sulphur components in atmospheric precipitation.

INTRODUCTION

The aim of this study is to determine the anthropogenic impact on weathering processes in the Carpathians. Environmental studies in the Carpathians are very important because of the presence of several National Parks (Babia Góra National Park, Gorce National Park, Magura National Park, and Bieszczady National Park in the Outer Carpathians, Pieniny National Park in the Pieniny Klipen Belt, and Tatra National Park in the Inner Carpathians). Some of these parks are part of international biosphere reservations, and numerous sandstone tors and picturesque rock groupings are also protected as nature reservations.[1]

Rock weathering in the presence of anthropogenic pollution of the atmosphere differs from natural weathering. The rate of weathering in a polluted atmosphere increases significantly in comparison with natural processes. The rate of weathering is related to the composition of rocks, their structure and to the concentration of aggressive gaseous components of air pollution and anthropogenic dust particles.[2,3] Visual manifestations of natural weathering and those accelerated by air pollution are, however, sometimes similar.

Several features can be considered as indicators of the influence of air pollution on the rocks in the study area:[4]

- The occurrence of compact, continuous gypsum crusts (or sometimes of other composition, e.g. other sulphates, halite) on the rock surface. The structure of the crust appears to be related to pollution concentration with dispersed single gypsum crystals occurring in areas of relatively low air pollution.
- The occurrence of a surface layer of amorphous or poorly crystalline aluminosilicates with a relatively high concentration of sulphur, chlorine, and phosphorus on the rock surface.
- The presence of numerous anthropogenic dust particles on the surface of the rocks. Not all anthropogenic dust particles can be easily distinguished from natural, aeolian material. Single particles of anthropogenic dust also occur in areas of low atmospheric pollution.
- A higher than natural concentration of some chemical elements, such as sulphur and chlorine, and heavy metals on rock surfaces or in the superficial layers of the rock. Different processes contribute to the concentration of some elements. Very often it is difficult to distinguish between those induced by atmospheric pollution and those associated with unpolluted conditions (e.g. The concentration of sulphur, phosphorous, and metals related to micro-organisms inhabiting the rock surface).

The presence of these visual manifestations of decay is not directly related to an increase in the weathering rate or to changes in weathering mechanisms. All values presented as the increase of rock weathering rate in polluted areas vary strongly in relation to rock type and local conditions. The weathering rate of limestone in urban or industrial areas can, for example, be slightly higher (1.35:1–1.47:1)[5] or significantly higher (5:1–7:1)[6,7] in comparison with rural areas.

Evaluation of the increase of weathering rates of sandstone is more difficult than is the case with limestone. In sandstone, components with different resistance to aggressive atmospheric parameters are present (e.g. resistant quartz grains or quartz cement, medium-resistant feldspars, and unresistant carbonate cement or carbonate grains). Some clay minerals, present in sandstone matrix, can act as sorbents of moisture and facilitate chemical reactions in the superficial layer of the rock.

SANDSTONES IN THE FLYSCH CARPATHIANS AND THEIR WEATHERING

The Flysch Carpathians (the Outer Carpathians) are mainly composed of turbidites, the sedimentation of which started in the Late Jurassic and continued until the Late Miocene. The main folding movements in the Outer Carpathians took place during the Miocene. The Polish part of the Carpathians comprises the northern fragments of the Inner Carpathians, the Pieniny Klipen Belt, and the Outer Carpathians (Figure 12.1). Numerous types of sandstone are present in the Outer Carpathians and these sandstones differ in grain-size, framework composition, degree of mineral and textural maturity, composition of cements and content of matrix. In addition they also exhibit different sedimentary structures and bedding characteristics.

The studied sandstones are represented by sub-lithic and lithic arenites or sub-arkoses. The matrix content is relatively low. Medium-grained sandstone samples characterized by a low content of carbonate cement were chosen for analysis. Samples were carefully studied to avoid samples containing pyrite, pyrite oxidation related sulphates, and barite.

Exposure to weathering factors varies due to the place and form of occurrence (outcrops in deep valleys, on mountain ridges, old quarries, isolated tors, abundance of vegetation, and microenvironment conditions). The manifestations of weathering on the Carpathian Flysch sandstones are numerous with granular disintegration and exfoliation being the most common. Formation of Fe-rich (or Fe- and Mn-oxide-rich) hardened layers on the rock surface also occurs in addition to honeycomb weathering and pit development.[8,9,10,11] The intensity of natural weathering is related to many factors which include lithological characteristics of the sandstone, local climatic conditions and development of vegetation cover.

The influence of anthropogenic air pollution on weathering processes in the Flysch Carpathians has not been assessed but

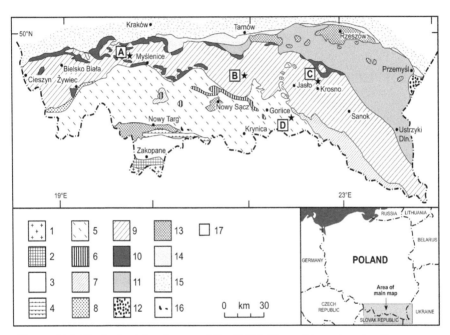

Figure 12.1 Geological sketch-map of the Polish Carpathians showing the structural units (after Malata, E., Malata, T. and Oszczypko N. (1996) Litho- and biostratigraphy of the Magura Nappe in the eastern part of the Beskid Wyspowy Range (Polish Western Carpathians). Annales Societatis Geologorum Poloniae 66: 269–284; partly modified). 1 – crystalline core of the Tatra Mts., 2 – High Tatra and sub-Tatra units, 3 – Podhale flysch, 4 – Pieniny Klippen Belt, 5 – Magura nappe, 6 – Grybów unit, 7 – Dukla unit, 8 – Fore Magura unit, 9 – Silesian unit, 10 – sub-Silesian unit, 11 – Skole unit, 12 – Stebnik unit, 13 – Miocene upon the Carpathians, 14 – Zgiobice unit, 15 – Miocene deposits of the Carpathian Foredeep, 16 – andesites, 17 – sample sites: A – Kalwaria Zebrzydowska, B – Ciezkowice, C – environs of Krosno (Odrzykon, Przadki), D – environs of Folusz (Diabli Kamien, Zamki nad Mrukowa), D – the Bieszczady Mts. (several points).

Alexandrowicz and Pawlikowski suggested that some gypsum-containing efflorescences on sandstones could be attributed to the elevated concentration of sulphur oxides in the atmosphere.[12] Wilczyńska-Michalik and Michalik presented a comparison of the weathering of sandstones subjected to the heavily polluted urban atmosphere in Kraków and those from the natural outcrops in the Carpathians.[13] Preliminary studies of the evidence of anthropogenic impact on the weathering of sandstones in various localities in the Carpathians indicate that the influence of atmospheric pollution, although generally low, varies significantly from site to site.[14,15]

ATMOSPHERIC POLLUTION IN THE CRAPATHIANS

The concentration of atmospheric pollution in the Carpathians differs significantly between western and eastern parts of the mountain chain in Poland. In the western sector the concentration of pollution is comparable with that from industrial areas and large urban agglomerations situated in close vicinity (e.g. Upper Silesia, Kraków). For example, the average influx of SO_4^{2-} is significantly higher in the western part of the Polish Flysch Carpathians (Figure 12.2).[16] While the influx of Cl^- is more than twice as high in the western sector (Figure 12.3).[17] The variation of pH values is not very high in the study area: in the western part of the Polish Outer Carpathians it is about half a unit lower than in the eastern part in the same measuring period (4.1–4.6 and 4.6–5.1 respectively).[18]

The differences in concentration and influx of atmospheric pollution are related to the regional airborne transport of pollutants from industrial and urban centres in Poland and partly from abroad, but more localized industrial and municipal emissions (both gaseous and dust) are also important sources. The variation of local emission levels in different administrative units in the Carpathians is very high.[19] In the former Bielsko-Biała voivodeship (western part of the studied area) total gaseous emission was about 1950×10^3 tonnes/year in 1996 and 731×10^3 tonnes/year in 1998 and in the former Krosno

INFLUENCE OF ANTHROPOGENIC FACTORS 231

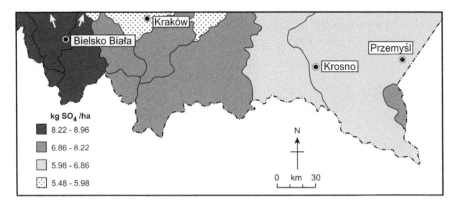

Figure 12.2 Influx of $SO_4{}^{2-}$ in the Carpathians (October–December 1998) (after Blachuta, J. and Twarowski, R. (1999) Applications and possibility of usage of spatial information in monitoring. 'Monitoring of chemical composition of atmospheric precipitation and deposition of pollution', Seminar: 19–29.; partly simplified).

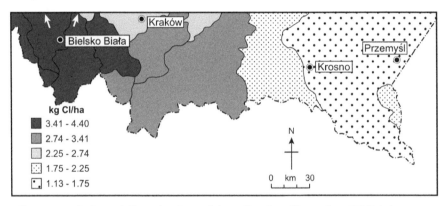

Figure 12.3 Influx of Cl^- in the Carpathians (October–December 1998) (after Blachuta, J. and Twarowski, R. (1999) Applications and possibility of usage of spatial information in monitoring. 'Monitoring of chemical composition of atmospheric precipitation and deposition of pollution', Seminar: 19–29.; partly simplified). Arrows – direction to the Upper Silesia industrial area.

voivodeship (eastern part) 54×10^3 tonnes/year in 1996 and 99×10^3 tonnes/year in 1998. The differences in the emission of atmospheric pollution are related also to differences in the density of population. In the Bielsko-Biała area the density is about 300 persons/km² and in the Beskid Niski Mts. and Bieszczady Mts (eastern part of the Polish Carpathians) below 60 persons/km².[20]

Table 12.1 Chemical composition of precipitation water in Kraków and in the Beskid Niski Mts. (November 1999)

Component	Sample	
	OP1199G (Kraków)	OP1199BIE (Beskid Niski Mts.)
Cl [mg/l]	24	<0.1
NO_3 [mg/l]	1	0.3
HCO_3 [mg/l]	24.4	12.2
SO_4 [mg/l]	31.5	0.2
PO_4 [mg/l]	0.064	0.005
Na [mg/l]	0.4	0.25
K [mg/l]	0.5	0.25
Ca [mg/l]	11.5	0.64
Mg [mg/l]	0.4	0.11
Fe [mg/l]	0.046	0.018
Mn [mg/l]	0.104	0.015

Differences between the chemical composition of precipitation in a one-month period (November) from urban (Kraków) and rural areas (Rzepedź, the Beskid Niski Mountains) are shown in Table 12.1.

SAMPLING SITES

The sampling sites for the present study have been chosen according to the following criteria:

♦ similarity in texture and composition of sandstones studied (medium-grained rocks with relatively low content of carbonate cement; devoid or almost devoid of pyrite),
♦ diversity in concentration of air pollution (dust and gases).

Different parts of the Beskidy Mountains (the northern part of the Beskid Makowski Mts., Beskid Niski Mts., Bieszczady Mts.) and the Carpathian Foothills (the Dynów Foothill) were chosen for sampling. About fifty samples were collected from surface layers of sandstone in natural outcrops and from buildings in which sandstone of local origin had been used as the building material. Twenty two samples were selected for detailed analysis.

LABORATORY ANALYSES

After preliminary selection based on field observation, samples were subjected to microscopic investigations. Both optical and electron microscopy were used. A scanning electron microscope fitted with an energy dispersive spectrometer (SEM-EDS) was used for the determination of the morphology and chemical composition of primary and secondary rock components. Natural surfaces of weathered rock and polished sections perpendicular to the surface were also examined. A carbon film coating of sample surfaces was applied before SEM-EDS analyses.

RESULTS

In Kalwaria Zebrzydowska (figure 12.1, point A), a relatively highly polluted area in the Carpathians, gypsum commonly occurs on rock surfaces in two different forms: compact, dark coloured gypsum crusts and dispersed, usually small gypsum crystals (invisible to the naked eye).

The dark coloured gypsum crusts occur on stone surfaces sheltered from direct rainwash. The crusts are composed of numerous platy gypsum crystals, which are more or less euhedral (Figure 12.4) with anthropogenic dust particles often present between the gypsum crystals (Figure 12.4). Blistering and exfoliation of the gypsum crust is relatively common and together with the detached gypsum crust

234 STONE DECAY ITS CAUSES AND CONTROLS

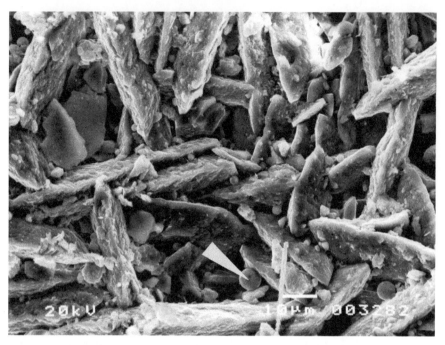

Figure 12.4 Gypsum crust on sandstone surface protected to rainwater washout (Kalwaria Zebrzydowska); between platy gypsum crystals spherical industrial dust particles (arrow). SEM.

some components of the sandstone are removed from the rock surface.

Dispersed, small (visible in SEM) gypsum crystals are present on surfaces exposed to the rainwash. Euhedral gypsum crystals form thick plates (Figure 12.5) although the voids observed in some of the gypsum crystals may be related to partial dissolution (Figure 12.6). The number of gypsum crystals varies significantly from place to place, but the surfaces on which these dispersed gypsum crystals occur more frequently exhibit relatively intense granular disintegration.

In the Carpathian foothills (Ciężkowice, point B and Krosno area, point C, Figure 12.1) and in the Beskid Niski (point D, Figure 12.1), where the concentration of air pollution is lower, continuous, compact gypsum crusts on rock surfaces have not been recorded. Only dispersed efflorescences and single gypsum crystals were found on rock surface or between detrital grains of sandstone (Figure 12.7).

INFLUENCE OF ANTHROPOGENIC FACTORS 235

Figure 12.5 Dispersed gypsum crystals on sandstone surface washed by rain water (Kalwaria Zebrzydowska). SEM.

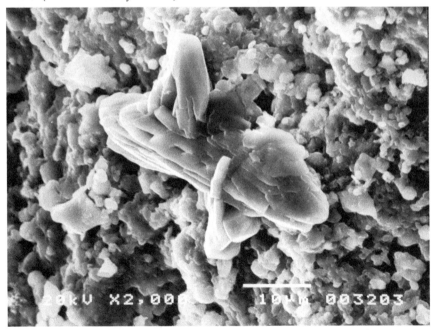

Figure 12.6 Gypsum crystals on sandstone surface (Kalwaria Zebrzydowska); voids related to dissolution. SEM.

Figure 12.7 Isolated group of gypsum crystals on sandstone surface growing from the space between detrital grains (Ciezkowice). SEM.

The observed variation in the amount of gypsum on rock surfaces of different samples is probably related more to local conditions (exposure of the rock surface and lithological characteristics) than to differences in the concentration of air pollution. In addition to gypsum, potassium alum was noted on the surface of the sample from the Ciężkowice area. In each of the above-mentioned sites, numerous anthropogenic dust particles were also observed on rock surfaces. Spherical aluminosilicate dust particles related to coal combustion in power plants occur most commonly.

SEM examination of samples from Odrzykoń near Krosno (the Carpathian foothills, point C, Figure 12.1) showed the corrosion and pitting of quartz grains and numerous aluminosilicate grains, barite and dolomite were observed (Figures 12.8–12.10) although the irregular grains of barite are very small (1–2 μm).

INFLUENCE OF ANTHROPOGENIC FACTORS 237

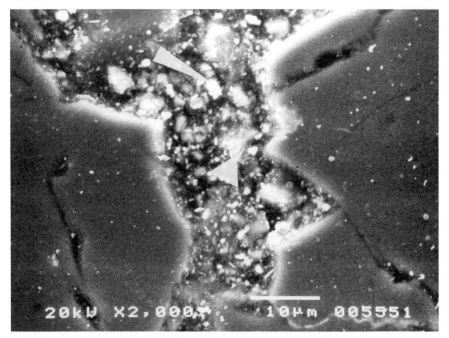

Figure 12.8 Embayments in quartz grains: numerous aluminosilicate grains, barite (long arrow; see EDS spectrum – Figure 12.9) and dolomite (short arrow; see EDS spectrum, Figure 12.10); (Odrzykoń near Krosno). SEM.

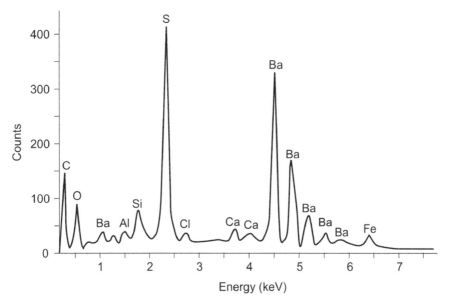

Figure 12.9 Energy dispersive spectrum of barite (see Figure 12.8).

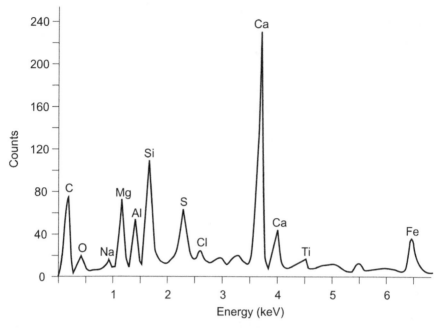

Figure 12.10 Energy dispersive spectrum of dolomite (see Figure 12.9).

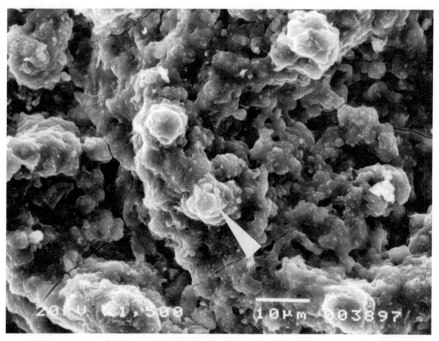

Figure 12.11 Irregular silica-rich layer (arrow – EDS analysis point; see Figure 12.12) (Ciezkowice). SEM.

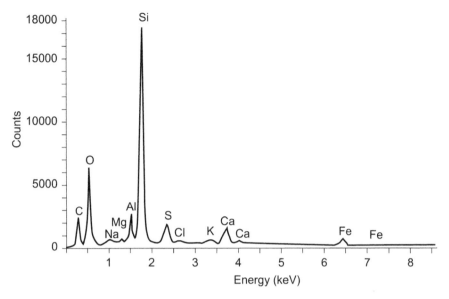

Figure 12.12 Energy dispersive spectrum of silica-rich layer (see Figure 12.11) on sandstone surface in Ciezkowice.

In the Bieszczady Mountains, where the concentration of air pollution is low in comparison with other localities, crystallization of salts on rock surfaces has not been observed and the presence of anthropogenic dust particles on rock surfaces are the only manifestations of atmospheric pollution.

Smooth or irregular layers composed mainly of silica ('silica skin'), but containing small amounts of aluminium, iron, calcium, sulphur, sodium, and potassium, cover surfaces of some samples from different localities and in some cases relatively high concentrations of sulphur occur (Figures 12.11–12.12).

DISCUSSION

The occurrence of gypsum crusts, dispersed efflorescences of gypsum, single gypsum crystals and the presence of anthropogenic dust particles on rock surfaces are related to atmospheric pollution. The composition of the rocks studied indicates that it is possible to exclude the influence of weathering of sulphides in sandstones on the development of sulphates in the studied samples.

The composition of atmospheric precipitation (Table 12.1) suggests that gypsum can be formed in the superficial layer of the rock by the reaction between SO_4^{2-} from atmospheric precipitation and Ca^{2+} derived from decomposed rock components especially calcite. The presence of voids in gypsum crystals indicates that rain is able to dissolve, at least temporarily, previously formed gypsum. Similar dissolution voids have been described in a rural area situated northwest of Kraków.[21]

Because of the porous nature of the sandstone, the gypsum forms within surface and near-surface pores as well as forming a surficial crust. The substrate material is weakened as a result of crystallization of gypsum within the pore spaces and when the crust becomes detached surface sandstone debris is lost with it.

The formation of gypsum crusts is limited to rock surfaces sheltered from rainwash. In the same locality on surfaces subject to direct washout by rainwater there are numerous small gypsum crystals dispersed on the surface and among framework grains in the superficial layer. The crystallization of gypsum (and/or other salts) in the superficial layer of the rock is associated with accelerated granular disintegration although the action of salt crystallization may be enhanced at high altitudes by frost.[22] The process of decay is comparable to the deterioration of sandstone observed in urban areas situated close to the study area,[23] to salt weathering in coastal areas[24] and in other natural environments.[25] In the sites examined evidence suggests that the salts identified are primarily derived from atmospheric pollution washed out in rainfall as they are not components of the sandstone nor of the groundwater. Although assessment of the acceleration of sandstone weathering arising from the addition of pollutants in the Carpathians is difficult, it may be that in some areas, where gypsum crusts have formed and crystallization of dispersed gypsum crystals is widespread, salt weathering induced by atmospheric pollution may be significant.

In areas of relatively low concentration of atmospheric pollution, where only dispersed efflorescences and single gypsum crystals on rock surface are noted, the role of this type of weathering does not appear to be so significant. However, it is possible that crystallization of gypsum between quartz grains in a superficial sandstone layer can over time modify the original rock texture by disrupting the granular framework thus facilitating the penetration of atmospheric precipitation and increasing susceptibility to weathering.

The presence of barite in the superficial layer of the sandstones indicates the presence of sulphate ions in the ambient atmosphere. The absence or scarcity of gypsum on rock surfaces from the same locality can be explained by the removal (dissolution) of gypsum from the surface by rains of relatively low pH and/or of low content of dissolved components (e.g. two orders lower concentration of SO_4^{2-} in the rural area of Beskid Niski than in Kraków, see Table 12.1). Barite is formed in the reaction between atmospheric SO_4^{2-} and Ba^{2+} ions originating, most probably, from decomposed K-feldspars. Newly formed barite is a relatively 'insoluble' mineral in the superficial rock layer and may be considered as evidence of the presence of sulphur compounds in rainwater.

In the Bieszczady Mountains (point E, Figure 12.1) gypsum or other salt crystals have not been noted. The only manifestation of atmospheric pollution is the presence of anthropogenic dust particles produced in fuel combustion.[26]

The formation of a thin silica-rich layer or 'silica skin' observed in numerous localities in the Carpathians occurs commonly on quartz-rich sandstones.[27] The process may be abiotic, but the activity of bacteria or lichen in the formation of silica flowstone is possible.[28,29] Lee and Parsons suggest that the silica-rich layer produced by lichen on granite may act as a 'protective' skin.[30] The formation of silica-rich layers is not related to anthropogenic air-pollution. However, the incorporation of sulphur into silica-rich layers and the formation of local concentrations of this element noted in several localities are related to its availability in the weathering environment in the Carpathians.

CONCLUSION

The presence of atmospheric pollution is manifest on rock surfaces in the Carpathians by the occurrence of gypsum crusts, gypsum efflorescences and particles of anthropogenic dust. However, rates and mechanisms of natural weathering do not appear to be significantly changed in most of the studied sites.

Only locally does salt weathering related to high concentrations of atmospheric pollution significantly accelerate rock decay. The presence of gypsum on the rock surface and within near-surface pores as well as forming a surficial crust is a typical feature of salt weathering. Salt weathering, significantly influenced by high levels of atmospheric pollution, has been noted only in a restricted area in the western part of the Polish Carpathians.

Rainwater of relatively low pH (noted over large areas of the Carpathians) and of low total concentration of dissolved components inhibit the formation of gypsum crusts and facilitate the dissolution of primary cements (e.g. carbonates) in the rock because of their undersaturation with respect to gypsum and calcite. Barite, with a lower solubility, is stable and its precipitation can be considered as the evidence for the presence of sulphur components in atmospheric precipitation. Local concentrations of sulphur in thin silica-rich layers on rock surfaces are also related to atmospheric pollution.

The data presented show that atmospheric pollution affects rock weathering in both natural outcrops and when used as a building material in urban environments close to industrialized areas. Over areas of the Carpathians characterized by low concentrations of air pollution, the natural weathering processes are not significantly modified.

References

1 Alexandrowicz, Z. (1996) Evaluation criteria in respect of the Carpathian geoconservation. In Alexandrowicz, Z. (ed.) Geoconservation of the Beskid Sadecki Mountains and the Sacz Basin, Polish Carpathians, *Studia Naturae* 42: 17–27.
2 Wilczyńska-Michalik, W. and Michalik, M. (1996) Weathering Processes of Sandstones from the Areas of Different Concentration of Air Pollution (in Polish, English summary). In: 1st International Conference on Theory and practice of atmospheric air protection, Polish Academy of Sciences, Institute of Environmental Engineering: 893–903.
3 Michalik, M. and Wilczyńska-Michalik, W. (1998) The influence of air pollution on weathering of building stones in Kraków. In Sulovsky, P. and Zeman, J. (eds.) ENVIWEATH 96, Environmental Aspects of Weathering Processes, *Folia Facultatis Scientiarum Naturalium Universitatis Masarici Brunnensis, Geologia* 39: 159–167.
4 Wilczyńska-Michalik, W. and Michalik, M. (1996) Petrographical, mineralogical and geochemical indicators of the impact of anthropogenic air-pollution on rocks (in Polish, English summary). *Rocznik Nauk.-Dydaktyczny WSP, Prace Geograficzne*, 16: 71–81.
5 Jaynes, S. and Cooke, R.U. (1987) Stone weathering in southeast England. *Atmospheric Environment* 21: 1601–1622.
6 Feddema, J.J. and Meierding, T.C. (1987) Marble weathering and air pollution in Philadelphia. *Atmospheric Environment* 21: 143–157.
7 Inkpen, R.J., Cooke, R.U. and Viles, H.A. (1994) Processes and rates of urban limestone weathering. In Robinson, D.A. and Williams, R.B.G. (eds.) *Rock weathering and Landform evolution.* John Wiley and Sons, Chichester: 119–130.
8 Alexandrowicz, Z. (1978) Sandstone tors of the Western Flysch Carpathians (In Polish; English summary). *Polska Akademia Nauk, Oddział w Krakowie, Komisja Nauk Geologicznych, Prace Geologiczne* 113: 6–87.
9 Alexandrowicz, Z. and Pawlikowski, M. (1982) Mineral crusts of the surface weathering zone of sandstone tors in the Polish Carpathians. *Mineralogia Polonica* 13: 41–59.
10 Alexandrowicz, Z. (1989) Evolution of weathering pits on sandstone tors in the Polish Carpathians. *Zeitschrift fur Geomorphologie N. F.* 33: 275–289.
11 Alexandrowicz, Z. and Brzeźniak, E. (1989) Dependences of weathering processes on surface sandstone rocks as a result of thermo-humidity changes in the Flysch Carpathians (In Polish). *Folia Geographica, Series Gepgraphica-Physica* 21: 17–36.
12 Alexandrowicz and Pawlikowski, *op. cit.* (1982)

13   Wilczyńska-Michalik, W. and Michalik, M. (1996) Weathering Processes of Sandstones from the Areas of Different Concentration of Air Pollution (in Polish, English summary). In: 1st International Conference on Theory and practice of atmospheric air protection, Polish Academy of Sciences, Institute of Environmental Engineering: 893–903.
14   Michalik, M. and Wilczyńska-Michalik, W. (1998) Sulphate minerals and anthropogenic dust particles on the surfaces of sandstones in the Carpathians as the indicators of atmospheric pollution concentration (In Polish, English summary). *Roczniki Bieszczadzkie* 7: 209–225.
15   Wilczyńska-Michalik, W. and Michalik, M. (1999) Weathering of the Carpathian flysch sandstones – a natural or atmospheric-pollution influenced process. *Geologica Carpathica* 50: 87–89.
16   Błachuta, J. and Twarowski, R. (1999) Applications and possibility of usage of spatial information in monitoring. 'Monitoring of chemical composition of atmospheric precipitation and deposition of pollution', Seminar: 19–29.
17   Twarowski, pers. comm.
18   Ibid.
19   Michalik, M. and Wilczyńska-Michalik, W. (1998) Sulphate minerals and anthropogenic dust particles on the surfaces of sandstones in the Carpathians as the indicators of atmospheric pollution concentration (In Polish, English summary). *Roczniki Bieszczadzkie* 7: 209–225.
20   Kondracki, *op. cit.* (1998)
21   Wilczyńska-Michalik, W. and Michalik, M. (1998) Differences of the mechanisms of weathering of the Jurassic limestones related to the concentration of air pollution. In Sulovsky, P. and Zeman, J. (eds.) ENVIWEATH 96, Environmental Aspects of Weathering Processes, *Folia Facultatis Scientiarum Naturalium Universitatis Masarici, Geologia,* 39: 233–239.
22   Robinson, D.A. (1981) Weathering of sandstone by the combined action of frost and salt. *Earth Surface Processes and Landforms* 6: 1–9.
23   Michalik, M. and Wilczyńska-Michalik, W. (1998) The influence of air pollution on weathering of building stones in Kraków. In Sulovsky, P. and Zeman, J. (eds.) ENVIWEATH 96, Environmental Aspects of Weathering Processes, *Folia Facultatis Scientiarum Naturalium Universitatis Masarici Brunnensis, Geologia* 39: 159–167.
24   Moropoulou, A. and Theoulakis, P. (1992) Conditions causing destructive NaCl crystallization into the porous sandstone, building material of the medieval city of Rhodes. In Decrouez, D., Chamay, J. and Zezza, F. (eds.) *The conservation of monuments in the Mediterranean Basin.* Proceedings of the 2nd International Symposium, Genève, 1991. Ville de Genève – Muséum d'Histoire Naturelle and Musée a'art. Et d'historie: 493–499.
25   Goudie, A. and Viles, H. (1997) *Salt Weathering Hazards.* John Wiley and Sons, Chichester.

26 Michalik, M. and Wilczyńska-Michalik, W. (1998) Sulphate minerals and anthropogenic dust particles on the surfaces of sandstones in the Carpathians as the indicators of atmospheric pollution concentration (In Polish, English summary). *Roczniki Bieszczadzkie* 7: 209–225.
27 Wray R.A.L. (1999) Opal and chalcedony speleothems on quartz sandstones in the Sydney region, southeastern Australia. *Australian Journal of Earth Sciences* 46: 623–632.
28 Barker, W.W., Welch, S.A., Chu, S. and Banfield, J.F. (1998) Experimental observations of the effects of bacteria on aluminosilicate weathering. *American Mineralogist* 83: 1551–1563.
29 Lee, M.R. and Parsons, I. (1999) Biomechanical and biochemical weathering of lichen-encrusted granite: textural controls on organic-mineral interaction and deposition of silica-rich layer. *Chemical Geology* 161: 385–397.
30 Ibid.

13 Observations on Stepkarren Formed on Limestone, Gypsum and Halite Terrains

D. MOTTERSHEAD and G. LUCAS

ABSTRACT

Small-scale step forms are described on exposed surfaces developed on limestone, gypsum and salt rock. They include both simple heelkarren and other apparently related forms for which the term stepkarren is preferred. An explanation of stepkarren formation is required which is capable of general application across lithological groups and climatic environments. A distinctive feature of the steps is the occurrence at the step surface of a microcrystalline layer not recognized by previous studies. A model of stepkarren development is proposed which interprets the significance of this layer. On limestone, lichen appear to have a role in the formation of the microcrystalline layer, creating features resembling lichen stromatolites.

INTRODUCTION

A common feature of small-scale karren landscapes on exposed surfaces of soluble rock is the presence at the scale of centimetres of distinctive freely drained slope facets of low gradient. Such features were first described in a systematic context by Bögli in 1960 as Trittkarren,[1] and have been subsequently observed by Sweeting (1972),[2] Werner (1975),[3] Vincent (1983),[4] and Choppy (1996)[5] on limestone terrains. Calaforra[6,7,8] and Macaluso and Sauro[9,10] have described them from gypsum terrains. Similar forms have also been studied in the context of limestone terrains in Mallorca by Smart and Whittaker who employ the term 'stepped flats',[11] and Crowther who describes them as 'steps'.[12]

A classification of small-scale solutional landforms is presented by Ford and Williams in 1989.[13] Curiously, gradient, a defining parameter of most erosional processes, does not feature as a diagnostic criterion. Within this classification the general form of freely drained flats is represented solely by Trittkarren (heelprints).

Ford and Williams describe trittkarren as characterized by an arcuate headwall, with a flat floor, normally 10–30 cm in diameter, and open at the downslope end.[14] Upslope, the heel is partially enclosed by a scarp some centimetres in height, which is often dissected by solution flutes; the heel may be enclosed laterally by quasi-linear divides. Bögli (1980) translates 'Trittkarren' as 'heel-print karren',[15] a term commonly adopted by writers in the English language.[16] Werner, however, translates 'Trittkarren' as 'stepkarren'.[17] In this paper we prefer the latter term, which we regard as more appropriate since low gradient appears to be a more fundamental defining characteristic of the landform than the arcuate planform. The term 'stepkarren', in avoiding any specification of planform, is less restrictive than heelkarren. It therefore permits the inclusion of a wider range of generally recognizable and apparently related forms of which the common feature is a smooth freely drained surface of low gradient. This need for a broader definition is implicitly recognized by Smart and Whittaker[18] and Crowther.[19]

In this paper we report field observations of stepkarren from limestone, gypsum and halite terrains, which were made in the context of a broader study of the surface morphology of soluble rocks.[20] The observations made offer the opportunity for original comment in two respects. First, they include new information on surface phenomena not previously reported in the context of karren. Secondly, in ranging across several lithologies, they provide a comparative perspective beyond the scope of previous single-site studies of heelkarren. New observations are made which allow evaluation of previous studies, and permit a new interpretation to be advanced.

The objectives of the present study may thus be specifically defined as to:

♦ demonstrate the existence of a range of stepkarren forms,
♦ review the geographical and lithological distribution of stepkarren,
♦ investigate commonality of stepkarren form across different lithologies,
♦ present new information on the subsurface structure of stepkarren surfaces,
♦ interpret accretion and dissolution processes at the stepkarren surface,
♦ propose a model of stepkarren development,
♦ consider timescales of stepkarren evolution.

PREVIOUS STUDIES

Ford and Williams state that Trittkarren are relatively rare forms, and occur on gently inclined or stepped bare limestone or dolomite surfaces.[21] Bögli notes that they are locally common at high altitudes in the European Alps.[22] Vincent suggests that they are common in bare karst environments, including arctic environments, above or beyond the treeline.[23] Choppy records them from the Dachstein district of the Austrian Alps at an altitude of 1900 m.[24] Sweeting records them from the humid temperate lowland environment of The Burren, County Clare, Eire.[25]

Ford and Williams suggest that, lithologically, they are limited to homogeneous, fine grained to aphanitic limestone, dolomite or marble.[26] This is supported by Vincent's field study on a marble outcrop,[27] and Jennings' observations of such forms in the marble mountains of New Zealand.[28] Observations from Mediterranean environments, however, suggest that the distribution of these forms may be considerably wider, as Calaforra [29,30] and Macaluso and Sauro[31,32] describe Trittkarren developed on gypsum outcrops in unglaciated terrains in Spain and Italy respectively. The authors are unaware of any such observations previously reported from salt terrains.

In respect of their origin, Bögli explained stepkarren on limestone as due to a local diminution of the dissolution process,[33] as corrosion of the step becomes inhibited by the development of a layer of carbonate-rich water in contact with the rock. Haserodt, on the other hand, regards stepkarren as a sub-nival feature and ascribes them to micro snowdrift formation.[34] White ascribes the development of stepkarren, although without substantiation, to sheetflow pulsing irregularly across the rock surface.[35] Calaforra ascribes the origin of Trittkarren on gypsum to lithological variation in the form of a vein, fracture or textural change.[36]

Vincent provides a model of heelkarren development on the basis of morphometric correlations, and points out that little is known about their morphology or genesis.[37] Crowther observes that stepkarren possess much lower surface roughness than adjacent flute surfaces.[38,39]

Previous studies of stepkarren are typically concerned with a single site or lithology. They tend to concentrate on the arcuate planform, and on the simple isolated heelkarren form. They have left unresolved such issues as the reasons why stepkarren are initiated, and why sloping surfaces on outcrops of soluble rocks resolve themselves into alternations of steep scarplets (often fluted) and stepkarren forms, rather than a steady uniform gradient. The mechanisms of formation previously proposed have been largely speculative, and based on observation of morphology alone.

METHODS

Field sampling

Observations of stepkarren were made at four principal locations in Mediterranean Spain (Table 13.1). The features observed possess a commonality of both scale and low gradient, the latter normally <10°. They may show considerable variation in planform and are commonly backed by a headscarp. All fulfil, however, the diagnostic criterion of freely drained rock slope facets of low gradient. The study sites embrace the three major groups of soluble sedimentary rocks. At each location karren terrains occur across areas commonly measured in hundreds of metres, to provide an extensive basis for observation. Given the basis of site selection, the nature of the sampling of stepkarren is adventitious rather than deliberative. The procedure adopted nevertheless leads to some original observations and provides a comparative perspective not available to previous studies.

Table 13.1 Study locations in Spain.

Site	Province	Location		Lithology
Lluc	Baleares	39.43N	2.54E	Conglomeratic limestone
Bobadilla	Andalucia	37.02N	4.44W	Gypsum
Vallada	Valencia	38.54N	0.41W	Gypsum
Cardona	Cataluña	41.55N	1.41E	Salt rock, salt waste

Laboratory procedures

Thin sections were prepared of samples bearing stepkarren surfaces. Salt and gypsum samples were sectioned using oil immersion; limestone samples were cut with water. None was impregnated. They were examine using a Leitz Laborlux12 PolS petrological microscope, with images captured by 35 mm camera or by JVC video camera. Samples for SEM were gold coated and examined in a Cambridge Instruments Stereoscan 250 scanning electron microscope with energy dispersal x-ray facility.

STEPKARREN FORM

Stepkarren occur in varied geomorphological and hydrological settings within local terrains in soluble bedrock. They represent just one component of a range of small-scale landforms which develop in such terrains, and they occur in identifiable morphological associations with other forms.

Morphological and scale differences were observed between bedrock terrains developed on limestone and rocksalt on the one hand, and gypsum terrains on the other. The limestone and salt outcrops in this study commonly comprise rock masses extending over several square metres and with 1–2 metres of relief. In such terrains, local topographic highs act as watersheds and a suite of distinctive subsidiary micro-landforms develops. The majority of the gypsum exposures, however, tend to occur in quasi-horizontal individual beds, commonly decimetres in thickness, which emerge laterally from slopes. They receive direct precipitation, and may also receive runoff from a slope above, but tend not to form water shedding divides.

Gradient

In this study a major diagnostic feature of stepkarren forms is considered to be the low gradient. The gradient, however, is not constant, and tends to decline along the axial length of the step toward the downslope margin. A single gradient value therefore needs to be treated with caution. A number of samples provide an indicative range of values of stepkarren gradient. On limestones at Lluc a total of 32 observations show a range of 2°–8°, strongly clustered in the range 4°–5°. Ten measurements on gypsum showed a range of 1°–8°. In contrast, rocksalt steps showed a range of 9°–12°, whereas well-developed stepkarren on salt waste had gradients in the range 4°–10°.

Morphological associations

Well-developed solutional terrains in limestone and salt develop a gross arrangement of slope forms similar in general terms to large-scale terrain. They form catchment basins separated by divides, and identifiable valley and channel systems, as on the conglomeratic limestone at Lluc (Figure 13.1a).

Within such terrains stepkarren appear to develop in several kinds of location. They may develop below steeper slopes dissected by solution flutes, situated at a divide, or on a steep section of a midslope. The fluted slope, as the flutes fade out, commonly grades down to meet the flat.

Stepkarren may develop within valleys that form chutes between steep divides. At Lluc there are many examples of valley forms whose long profile comprises a sequence of alternate scarps and steps (Figure 13.1b). At Cardona a valley is formed in rocksalt in which the flat floor directly meets the steep bounding fluted slopes. In this case the stepkarren form extends without interruption across the marl bands interbedded with the rocksalt.

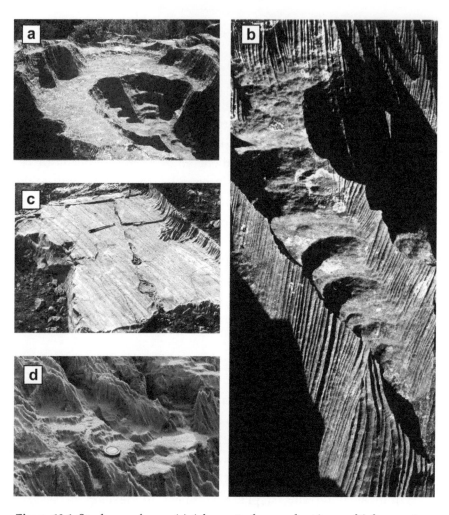

Figure 13.1 Stepkarren forms: (a) A large stepkarren showing multiple arcuate planforms at both upslope and downslope margins (Lluc) (b) A series of steps forming a drainage chute between divides (Lluc) (c) A broad step cut into an emerging bed of gypsum (Bobadilla). The upslope margin is fluted; the very steep downslope margin is microrilled. The step crosses uniformly across bedding structures, and joints and fissures (d) Multiple stepped slope on salt waste (Cardona). The original slope formed by tipping salt waste is presumed to be planar in profile; the slope has now resolved into an alternation of stepkarren and fluted scarps.

In gypsum terrains, a horizontal gypsum bed emerging from a slope will commonly possess a steep scarp sculpted into solution flutes (Figure 13.1c). These then fade out downslope and grade smoothly into a stepkarren surface. On salt waste, which is assumed to be of uniform structure and initial slope form, slopes commonly show an alternation of fluted scarps and stepkarren, arranged in the form of a staircase (Figure 13.1d).

A common feature of stepkarren form is that at their upslope and lateral margins, they are bounded by steeper slopes, commonly fluted, which grade down to the step surface. In contrast, the downslope edge of the stepkarren is much sharper, as the step is truncated downslope by a steep slope segment.

Planform

In the simple case stepkarren may adopt the planform of a single heel as commonly described by previous authors.[40,41,42] In this study, however, this simple form was not often seen. The planform of flats showed considerable variety, ranging from complex arcuate forms to linear steps transverse to the slope.

Where stepkarren form in catchment basins with arcuate watersheds, they tend to acquire an arcuate form. If the watershed has multiple arcuate forms, with several catchment heads feeding on to a composite stepkarren, the stepkarren too will acquire a similar planform. The downslope margin may be dissected by a single or multiple arcuate headcut(s), and thus acquire either a single arcuate or scalloped form. In this way the planform of the stepkarren may become quite complex. One such form observed at Lluc has a maximum width of 170 cm and a maximum length of 58 cm, with its leading edge cut by six scallop-form headcuts (Figure 13.1a). Where stepkarren form across a plane slope, as is often the case at the Salinera, Cardona, then they may form a linear staircase more or less straight in planform across the slope (Figure 13.1d).

The axial length of stepkarren is variable, but commonly in the range 80–150 mm.

Surface texture

The surfaces of stepkarren are characteristically smooth and possess substantially less surface roughness than adjacent fluted surfaces. This distinction exists at a scale visible to the naked eye, and also to the touch. The limestone stepkarren at Lluc have an almost polished feel in contrast to the rough, granular texture of nearby fluted surfaces. This distinction is confirmed morphometrically by Crowther.[43,44]

Surface microtopography at finer scales of resolution can be examined by scanning electron microscopy, and are illustrated by the following examples. A limestone step for Lluc possesses the topography of a mammilated plateau, with accordant rounded summits separated by pits, $c.10$ µm in diameter (Figure 13.2a). This kind of topography is apparently restricted to limestone, and appears to be associated with organic activity. An example of a gypsum flat shows, at $c.\times100$ magnification, the surface to be planed apparently indiscriminately across crystals from 50–350 µm in size (Figure 13.2b). A maximum relief of $c.40$ µm is apparent. At magnifications of $c.\times1000$,

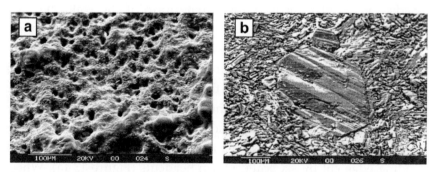

Figure 13.2 (a) An SEM image of mammilated plateau terrain forming the surface of a limestone step. The step surface is penetrated by circular pits of organic origin (×220) (b) A gypsum step surface showing is general smoothness, formed by an apparently planed surface of a large crystal, set in a microcrystalline matrix (×140).

the surface planed across larger crystals shows etching of their lattice structure, whereas areas of small crystals show inter-crystal boundaries to be etched to form crevices up to 10 μm in depth. On salt waste (not illustrated) polyhedral crystals show negligible relief with lattice structural lines picked out by troughs <5 μm in depth. The inter-crystal boundaries are marked by small crevices of <5 μm. In summary, detailed observations of stepkarren surfaces reveals very low relief at the microscale.

SUBSURFACE STRUCTURE

Nineteen thin sections were prepared from samples of the step surfaces (11 limestone, 5 gypsum, 3 salt). These enabled the surface form and subsurface structure of steps to be observed at the microscopic scale. Sample observations representative of each rock type are now presented, which demonstrate the range of features observed.

Limestone

Figure 13.3 (Lluc) shows the host rock to be a fine-grained microbreccia. Angular fragments of micrite, commonly 0.5 mm but up to 1–2 mm, are set in a sparite matrix of calcite crystals commonly 0.25 mm in size. The step surface is formed by a very fine-grained microcrystalline layer, comprising crystals of calcite sized 0.01–0.03 mm whose orientation normal to the topographic surface is picked out by impurities in the fabric. The lower boundary of the microcrystalline layer is irregular and reticulate as it follows the margins of the micritic texture of the host rock at a very fine scale. This boundary is marked by a darker brown layer, thought to be organic impurities. The thickness of the microcrystalline layer varies in the range 0.10–0.25 mm, representing the depth to which its accumulation has modified the topography of the step surface. The microcrystalline layer infills this irregular topography to form a smooth

258 STONE DECAY ITS CAUSES AND CONTROLS

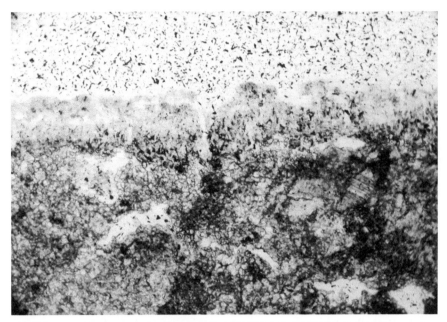

Figure 13.3 Thin section through a limestone step surface, clearly contrasting the structure of the microcrystalline layer with that of the host rock beneath (×100).

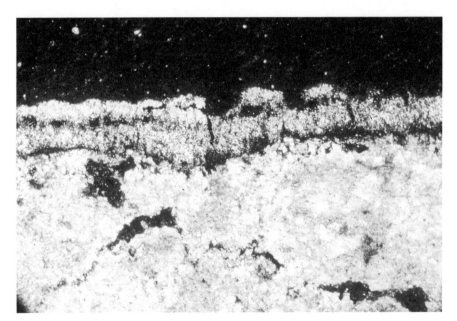

Figure 13.4 Thin section showing double microcrystalline layer on a limestone step. The darker layers represent concentrations of diffuse organic material (×40).

with a maximum local relief of $c.0.02$ mm. The surface layer appears to comprise a combination of both crystalline and presumed organic material. The latter does not possess any clear coherent form that might suggest a living organism.

Figure 13.4 shows a double layer of microcrystalline material infilling hollows in the host rock beneath. The host rock surface has a topographic amplitude of 0.10–0.25 mm, with highs at a lateral spacing of 0.5–1 mm. Topographic highs in the host rock surface appear to be formed by more coherent and therefore resistant crystal masses. Two hollows in this host rock surface relief are infilled by the fine crystalline layer, whose upper surface shows a maximum local relief of $c.0.02$ mm. The base of each stratigraphic unit of the microcrystalline material is marked by a darker layer, with upper fill nested within the lower one.

Figure 13.5 shows a step surface formed by a mixture of crystalline and organic material, and with a series of cavernous voids

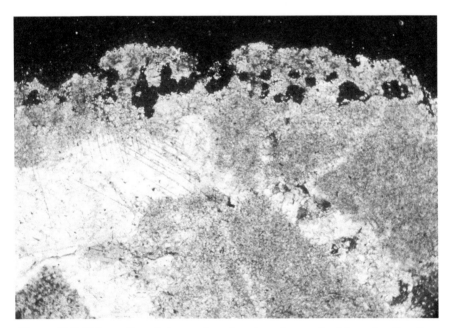

Figure 13.5 Thin section of step surface showing a layer of voids perforating the subsurface, characteristic of lichen growth, and the intimate mixture of crystalline and organic material. (×100).

 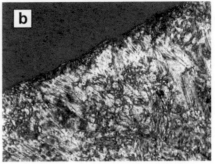

Figure 13.6 Two thin sections showing the microcrystalline layer forming the surface of a gypsum step (×100).

present at a depth of c.0.06 mm, resembling the perforations of a sheet of postage stamps. It strongly resembles sections through lichens published by Klappa, in which the voids represented the site of subcutaneous ascocarps.[45]

Gypsum

Figure 13.6 (a and b) shows a layer, 0.01–0.25 mm deep, of small gypsum crystals present at the surface of the step. This layer is visible in plane polarized light, and is separate from the main body of the gypsum host rock. It is shown by its differing birefringence colours and extinction angle of its component fabric, and is highlighted by the use of the fast plate.

In profile the surface of the underlying host rock beneath has a topographic relief of amplitude up to 0.25 mm and wavelength 1 mm. The contact between the surface layer and the host rock is irregularly reticulate around the margins of the partly etched gypsum crystals of the host rock. The microcrystalline layer infills this irregular topography to form a smooth surface topography at the step surface with a maximum local relief of c.0.02 mm.

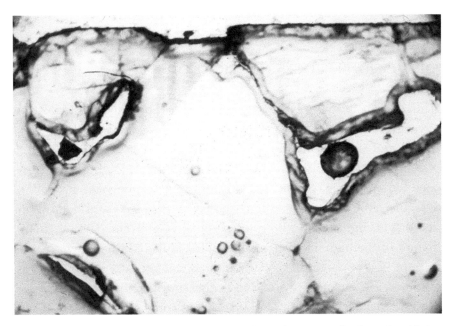

Figure 13.7 Thin section showing planed surface of step in rocksalt, and voids in the material beneath. Epitaxial crystal growth is visible both at the step surface and within the voids.

Salt

Figure 13.7 shows the host rock formed of halite crystals commonly exceeding 2 mm. The surface of the salt step is remarkably plane, with a relief of less than 0.001 mm except where it cuts across a crystal boundary. Epitaxial crystal growth is present on the step surface. No evidence has yet been found of a separate crystalline layer forming the surface of a step developed in salt.

A feature of this rock is the presence between the crystals of substantial voids, whose rounded outlines stand in strong contrast to the linearity of the crystal boundaries. Evidence of epitaxial crystal growth is present also around these voids, and along opposed crystal faces at some crystal boundaries.

DISCUSSION – PROCESS IMPLICATIONS

Thin sections provide evidence of the action of both dissolutional and crystallization processes at the surface of the soluble rocks in this study, and also beneath the surface.

Surface solution is shown by the irregularity of the host rock surface where it is preserved beneath the microcrystalline layer. Irregularity at the larger scale is shown by the peaks and hollows at the 0.1–1.0 mm scale in both limestone and gypsum. The reticulate nature of the boundary at the crystalline scale, clearly shown in the case of limestone, implies solution at crystal boundaries. In the case of limestone, subsurface solution is also shown to be effected by the action of lichen in creating voids in the microcrystalline layer. In the case of salt, subsurface solution is demonstrated both at planar crystal boundaries, and in the form of ovoid cavities, reminiscent of karstic caverns. Structural control is evident in all these cases, with resistance being demonstrated by larger and more mineralogically resistant crystals, and weakness by crystal boundaries.

Accretion by crystallization is shown by the accumulation of a microcrystalline layer, on both limestone and gypsum. The microcrystalline layer creates the smooth step surface at two scales; first, by the infilling of irregularities in the surface of the host bedrock and secondly by the crystallographic fineness of its own fabric. In the case of limestone, surface accretion may be accentuated by the presence of lichen, creating laminar accretions reminiscent of those described by Klappa as 'lichen stromatolites'.[46] In this way the accumulation of calcite, organic sedimentary material and the lichens themselves create laminae of mineral-rich and organic-rich layers. In the case of salt, subsurface crystallization in voids and at crystal boundaries is shown to occur.

This evidence suggests that, in response to wetting events, a complex of processes operates at and beneath the surface of soluble rocks. It follows from this that previous models of dissolutional erosion of soluble rock surfaces stand in need of reconsideration. The presence of a thin microcrystalline layer forming the surface of stepkarren has been demonstrated in respect of examples from both limestone and gypsum. On salt this layer may be represented by epitaxial crystal growth. Observation and inference suggest that it may both form and decay, forming by crystallization and decaying through solution. It may comprise single or multiple layers. The occurrence of this microcrystalline layer is a newly identified feature of stepkarren, and may be interpreted as integral to their formation and development.

The implications of the microcrystalline layer will now be discussed in respect of stepkarren forming processes. It appears that the layer is formed by evaporation of a mineral solution following a wetting event. Water would remain on the step surface, retained on the low gradient by microtopographic irregularities and surface tension. If it is assumed that the step surface is fully wetted by a precipitation event, then a fixed volume of water may be expected to remain. Provided that evaporation were able to run its full course, then a fixed amount of crystallization would occur, depending on the volume of water retained, and the concentration of dissolved minerals within it. In this case, assuming a known concentration of dissolved minerals, then a predictable depth of crystallization may be expected from such an event. Any wetting event may thus be expected to create a subsequent crystalline layer.

During the next precipitation event, renewed solution of the step surface will first attack the most recently crystallized surface layer. The extent to which this will be dissolved away will depend on the duration and magnitude of the precipitation event. A short event may not completely remove the crystalline layer. In this case, the subsequent evaporation phase would be expected to add a further increment to the remaining original layer. With a succession of such minor events, incremental growth of the crystalline layer may be expected. A long event however, with a substantial duration of overland flow, may be expected to remove entirely the previously formed microcrystalline layer and go on to attack the host bedrock beneath, further eroding it by dissolution, and thereby lowering the initial step surface. A subsequent phase would then deposit another new crystalline increment. Whether this survives then depends on the magnitude of succeeding events. Stochastic effects of magnitude and frequency therefore, would appear to play a crucial role in the continuing development or destruction and renewal of the microcrystalline layer.

A MODEL OF STEPKARREN DEVELOPMENT

Recognition of the microcrystalline layer now enables the postulation of a model of stepkarren development. Initiation would appear to be facilitated by the existence of a facet capable of retaining surface water. Figure 13.8 presents a model showing two possible sequences of development. At the initial stage of the next wetting event, erosion of the host bedrock beneath the step would be inhibited by the presence of the crystalline layer, unless and until the latter is removed. Thus step lowering would be initially inhibited by the microcrystalline layer. Further lowering of the subjacent bedrock would require the delivery of sufficient solvent to remove the microcrystalline layer. In contrast, the adjacent steep bounding slope above the step, where the host rock is directly exposed, is immediately subject to solutional erosion, and would suffer slope retreat. The base of this bounding slope, however, is constrained in elevation by the existence of the microcrystalline layer for as long as the latter survives. If the wetting event is of such magnitude that the layer is completely eroded, then erosion and lowering of the step surface will take place (Figure 13.8a). Alternatively, the wetting event may be insufficient to completely erode the microcrystalline layer, in which case the following evaporation phases may cause the continuing accretion of additional microcrystalline material (Figure 13.8b).

In this way, as long as the microcrystalline layer remains in place, whether it survives a wetting event or not, the flat would be extended incrementally headwards following each cycle of wetting and crystallization. Over time, the flat would extend by episodic

a

(i) Microcrystalline layer emplaced following evaporation event.

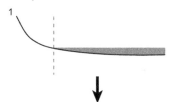

(ii) During new wetting event, bounding slope retreats headwards, microcrystalline layer protects host rock surface beneath step.

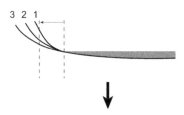

(iii) Microcrystalline layer eroded, both bounding slope and step now erode.

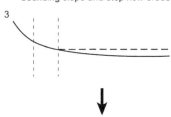

(iv) New microcrystalline layer emplaced on extended step.

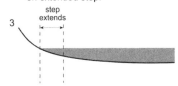

b

(i) Microcrystalline layer emplaced following evaporation event.

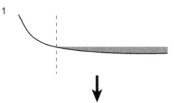

(ii) During new wetting event, bounding slope retreats headwards, microcrystalline layer protects host rock surface beneath step.

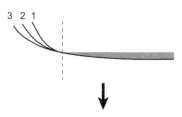

(iii) Microcrystalline layer thinned by erosion but not removed.

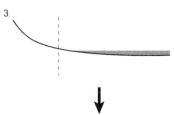

(iv) Accretion of further microcrystalline material deepens the layer

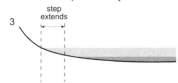

Figure 13.8 A model of stepkarren development, showing two possible modes.
(a) Episodic formation and erosion of the microcrystalline layer.
(b) Thinning and continuing accretion of the microcrystalline layer.

growth, and thus increase its capacity for retaining a water film during the next evaporation phase.

Thus the headslope is able to retreat continuously, whilst the step extends incrementally headwards causing the step to grow. It is reasonable to infer a general parallel retreat of the bounding slope.[47,48] The lower part of this slope, however, may be subject to less aggressive solution as a consequence of increasing solute load downslope, the onset of laminar flow, or the increasing depth of flow.[49] These effects may initiate a concavity in the lower slope profile,[50] and may therefore be explained in terms of chemical and mechanical mechanisms of local erosion (see Mottershead and Lucas, Chapter 14).

The step is terminated at its lower boundary by a further steep bedrock slope. This exposes the host bedrock and is therefore subject to continuous retreat throughout wetting events. Thus it is likely that in profile the entire stepped landscape comprises retreating host rock slopes, separated by steps whose location migrates over time as a consequence of the retreat of the upper and lower bounding slopes.

Vincent proposes a geometric model of heelkarren evolution through both headward and lateral retreat of the arcuate headwall slopes.[51] Although it deals with concurrent processes, the model may also be employed in the present context as a predictive tool, in order to identify the development of karren form over the longer term. Over a longer timescale it may reasonably be inferred that the laterally extending arcuate headwalls of adjacent steps will tend to intersect each other, leading to the development of a multicuspate planform of the arcuate upper bounding slopes. Headward erosion by the next heelkarren downslope would lead to undercutting, in which the trailing edge becomes nibbled out by the headscarp of the next lower step, thereby creating a yet more complex step planform. If this interpretation is correct, then the hitherto frequently observed fields of simple separate heelkarren forms would appear to represent a juvenile form, an initial stage of solutional denudation of a quasi-planar bedrock slope. The multiple stepped topography of Lluc and Cardona may represent the more complex forms of a more evolved micro-landscape of solutional erosion.

RATES OF DEVELOPMENT

The concept of increasing complexity through time in the evolution of heelkarren and stepkarren, apparently predicted by the Vincent model, may be tested by establishing a measure of the extent of micro-landscape evolution. A means of comparing exposed bedrock surfaces in this way may be developed by combining rates of surface erosion with the age of the surface. In this way the product of a general lowering rate and surface age will indicate the extent to which the landscape has evolved in terms of an estimated overall surface lowering value. Appropriate data are available for some exemplar sites, and are summarized in Table 13.2.

Table 13.2 Evolution rates of stepkarren-bearing terrains.

Environment	Lithology	Surface age (a^{-1})	Erosion rate (mma^{-1})	Total lowering (mm)	Stepkarren
Subarctic[a,b]	Marble	9500	0.005	50	Heelkarren only
Temperate maritime[c]	Limestone	12000	0.012	150	Heelkarren Incipient steps
Muntanya[d]	Salt waste	8	25–30	200–250	Incipient steps
Salinera[e]	Salt waste	20	20	400	Steps present

Sources

a André, M-F. (1996) Rock weathering rates in arctic and subarctic environments (Abisko Mts., Swedish Lappland). *Zeitschrift für Geomorphologie N.F.* 40: 499–517.
b P. Worsley, *pers. Comm.*
c Williams, P.W. (1966) Limestone pavements with special reference to western Ireland. *Transactions of the Institute of British Geographers* 40: 155–172.
d Mottershead, D.N., Moses, C.A. and Lucas, G.R. (2000) 'Lithological control of solution flute form: a comparative study', *Zeitschrift für*
e *Geomorphologie N.F.* 44: 491–512.
f Ibid.

The age of deglaciation of the surface on which Vincent observed heelkarren in arctic Norway is $c.9.5$ ka BP.[52] André suggests a weathering rate of 5 mmka^{-1} for marble in this subarctic environment;[53] this is equivalent to a general surface lowering of $c.50$ mm for this surface now bearing simple heelkarren. Williams postulates that limestone surfaces in the west of Ireland, which now bear heelkarren, have been lowered by 150 mm in postglacial time.[54] At the Muntanya (Cardona) site, a salt rock terrain some 8 years in age is exposed to solutional erosion at a rate of 25–30 mma^{-1}.[55] This terrain has undergone a general surface lowering of 200–250 mm, and stepped terrain is only poorly developed. At the nearby Salinera site, however, the authors have observed very well developed stepkarren on salt waste, on surfaces some 20 years old undergoing solutional erosion at a rate of 20 mma^{-1}. This surface has therefore undergone a general lowering of some 400 mm.

This limited evidence suggests that while a general surface lowering of 50 mm may be sufficient for the development of heelkarren, the development of a stepped landscape appears to require a minimum of at least 300 mm of general surface lowering. This is at least consistent with the hypothesis that simple heelkarren may be a juvenile form, and that their continuing development may be inferred to lead to more complex stepped terrains of the kind observed at Lluc.

The surfaces on which the complex stepkarren landscapes of Lluc are developed are likely to have evolved at a significant rate. Thus Cucci *et al.* have evaluated solutional loss of limestone in a Mediterranean environment near Trieste, under a comparable precipitation conditions of 1350 mma^{-1}, at 0.01–0.03 mma^{-1}.[56] The Lluc surfaces are of unknown age but are likely to have been continuously exposed to erosion for an unquantifiably long time.

CONCLUSIONS

The evidence presented in this study appears to support a number of conclusions worthy of further investigation.

- That currently accepted classifications and descriptive terms inadequately represent the range of forms observed in this study.
- That stepkarren are not restricted to limestones, and may develop on each of the major groups of soluble rocks; they therefore require an explanation of general application across different lithological groups and climatic environments.
- That stepkarren are not restricted to nival or recently glaciated environments; explanations requiring glacial[57] or nival[58] conditions are inadequate to account for occurrences in Mediterranean terrains.
- That previous inferential models of stepkarren formation by dissolutional processes alone now require revision.[59,60]
- That stepkarren may possess a more complex planform than the simple heelkarren; they clearly represent an integral morphological component of the solutional landscape system on exposed bedrock surfaces, and may well represent a more evolved form of heelkarren predicted by the model of Vincent.[61]
- That stepkarren possess a notably smooth surface, formed by the accretion of a microcrystalline layer, which is here reported for the first time.
- That biological processes may contribute to stepkarren formation in limestone terrains; the similarity of general form across limestone, gypsum and salt, however, implies the fundamental processes of stepkarren development are independent of lichen influence.
- That solutional erosion of low gradient surfaces in soluble rock materials may occur both at the surface and beneath it as microkarst forms develop at the sub millimetre scale.
- That the initiation and extension of stepkarren provide a mechanism for the explanation of stepped slopes on apparently homogeneous soluble rocks.
- That development of the microcrystalline layer may cause stepkarren to extend headward and laterally over time, leading to a complex stepped micro-landscape, typified by the stepped terrains of the Serra de Tramuntana, Mallorca.

Acknowledgements

Norman Jenkinson provided very helpful assistance with SEM. Harry Williams of Manchester University prepared thin sections for analysis. Jose Maria Calaforra and Felix Teixidor provided local information on the Spanish sites. Kathy Coffey prepared the thin sections. Ann Chapman created the artwork. Peter Stein provided a range of technical services. Dave Groom provided technical assistance in both field and laboratory. Edge Hill College of Higher Education contributed towards field expenses and analytical procedures. We are grateful to Dr. H.A. Viles who drew our attention to the work of C.F. Klappa.

References

1. Bögli, A. (1960) Kalklösung und Karrenbildung. *Zeitschrift für Geomorphologie Supplement* 2: 4–21.
2. Sweeting, M.M. (1972) *Karst landforms*. Macmillan, London.
3. Werner, E. (1975) Solution of calcium carbonate and the formation of Karren. *Cave Geology* 1: 3–28.
4. Vincent, P.J. (1983) The morphology and morphometry of some arctic Trittkarren. *Zeitschrift für Geomorphologie N.F.* 27: 205–222.
5. Choppy, J. (1996) Les cannelures et rigoles sont des indicateurs climatiques (karst profond et karst superficiel. In Fornós, J.-J., and Ginès, A. (eds) *Karren Landforms*. IUCN/IGU, Palma: 137–148
6. Calaforra, J.M. (1996) *Contribucion al conocimiento de la karstologia de yesos*. Universidad de Almeria, Almeria.
7. Calaforra, J.M. (1996) Some examples of gypsum Karren. In Fornós, J.-J. and Ginès, A. (eds) *Karren Landforms*. IUCN/IGU, Palma: 253–260.
8. Calaforra, J.M. (1998) *Karstologia de yesos*. Universidad de Almeria, Almeria: 384.
9. Macaluso, T. and Sauro, U. (1996) The Karren in evaporitic rocks: A proposal of classification. In Fornós, J.-J. and Ginès, A. (eds) *Karren Landforms*. IUCN/IGU, Palma: 277–293.
10. Macaluso, T. and Sauro, U. (1996) Weathering crust and Karren on exposed gypsum surfaces. *International Journal of Speleology* 25: 115–126.
11. Smart, P.L. and Whittaker, F.F. (1996) Development of karren landform assemblages – a case study from Son Marc, Mallorca. In Fornós, J.-J. and Ginès, A. (eds) *Karren Landforms*. IUCN/IGU, Palma: 111–122.

12 Crowther, J. (1997) Surface roughness and the evolution of Karren forms at Lluc, Serra de Tramuntana, Mallorca. *Zeitschrift für Geomorphologie N.F.* 41: 393–407.
13 Ford D.C. and Williams, P. (1989) *Karst geomorphology and hydrology.* Unwin Hyman, London.
14 Ibid.
15 Bögli, A. (1980) *Karst hydrology and physical speleology.* Springer-Verlag, Berlin.
16 Ford and Williams, *op. cit.* (1989)
17 Werner, *op. cit.* (1975)
18 Smart and Whittaker, *op. cit.* (1996)
19 Crowther, *op. cit.* (1997)
20 Mottershead, D.N., Moses, C.A. and Lucas, G.R. (2000) 'Lithological control of solution flute form: a comparative study', *Zeitschrift für Geomorphologie N.F.* 44: 491–512.
21 Ford and Williams, *op. cit.* (1989)
22 Bögli, *op. cit.* (1960)
23 Vincent, *op. cit.* (1983
24 Choppy, *op. cit.* (1996)
25 Sweeting, *op. cit.* (1972)
26 Ford and Williams, *op. cit.* (1989)
27 Vincent's, *op. cit.* (1983)
28 Jennings, J.N. (1985) *Karst geomorphology.* Blackwell, Oxford.
29 Calaforra, J.M. (1996) *Contribucion al conocimiento de la karstologia de yesos.* Universidad de Almeria, Almeria.
30 Calaforra, J.M. (1996) Some examples of gypsum Karren. In Fornós, J.-J. and Ginès, A. (eds) *Karren Landforms.* IUCN/IGU, Palma: 253–260.
31 Macaluso, T. and Sauro, U. (1996) The Karren in evaporitic rocks: A proposal of classification. In Fornós, J.-J.and Ginès, A. (eds) *Karren Landforms.* IUCN/IGU, Palma: 277–293.
32 Macaluso, T. and Sauro, U. (1996) Weathering crust and Karren on exposed gypsum surfaces. *International Journal of Speleology* 25: 115–126.
33 Bögli *op. cit.* (1960)
34 Haserodt, K. (1965) Untersuchungen zur Hohen- und Altersgliederung der Karstformen in den Nordlichen Kalkalpen. *Munchner Geographische Handbuch* 27.
35 White, W.B. (1988) *Geomorphology and hydrology of karst terrains.* Oxford University Press, Oxford.
36 Calaforra, J.M. (1996) *Contribucion al conocimiento de la karstologia de yesos.* Universidad de Almeria, Almeria.
37 Vincent *op. cit.* (1983)
38 Crowther, J. (1996) Roughness (mm-scale) of limestone surfaces: examples from coastal and sub-aerial Karren features in Mallorca. In Fornós, J.-J. and Ginès, A. (eds) *Karren Landforms.* IUCN/IGU, Palma: 149–159.
39 Crowther, *op. cit.* (1997)

40 Bögli, *op. cit.* (1960)
41 Sweeting, *op. cit.* (1972)
42 Vincent, *op. cit.* (1983)
43 Crowther, *op. cit.* (1996)
44 Crowther, *op. cit.* (1997)
45 Klappa, C.F. (1979) Lichen stromatolites: criterion for subaerial exposure and a mechanism for the formation of laminar calcretes (caliche). *Journal of Sedimentary Petrology* 49: 387–400.
46 Ibid.
47 Mottershead, D.N. (1996) A study of solution flutes (Rillenkarren) at Lluc, Mallorca. *Zeitschrift für Geomorphologie N.F.Supplement* 103: 215–241.
48 Crowther, *op. cit.* (1997)
49 Glew and Ford, *op. cit.* (1980)
50 Vincent, *op. cit.* (1983)
51 Ibid.
52 P. Worsley, pers. comm.
53 André, M-F. (1996) Rock weathering rates in arctic and subarctic environments (Abisko Mts., Swedish Lappland). *Zeitschrift für Geomorphologie N.F.* 40: 499–517.
54 Williams, P.W. (1966) Limestone pavements with special reference to western Ireland. *Transactions of the Institute of British Geographers* 40: 155–172.
55 Mottershead, D. and Lucas, G. (2001) Field testing of Glew and Ford's model of solution flute evolution. *Earth Surface Processes and Landforms* 26: 839–846.
56 Cucci, F., Forti, P. and Marinetti, E. (1996) Surface degradation of carbonate rocks in the karst of Trieste, Classical Karst, Italy. In Fornós, J.-J. and Ginès, A. (eds) *Karren Landforms*. IUCN/IGU, Palma: 41–51.
57 Sweeting, *op. cit.* (1972)
58 Haserodt, *op. cit.* (1965)
59 Bögli, *op. cit.* (1950)
60 White, *op. cit.* (1988)
61 Vincent, *op. cit.* (1983)

14 The Role of Mechanical and Biotic Processes in Solution Flute Development

D. MOTTERSHEAD and G. LUCAS

ABSTRACT

Previous models of solution flute formation explain their development in terms of hydrochemical considerations alone. Recent investigations have shown that mechanical and biotic processes also play an important part in flute forming processes. Evidence is presented to demonstrate the effects of these processes, and the need to incorporate them in models of flute development.

INTRODUCTION

Solution flutes are surface erosional forms which develop on soluble rocks directly exposed to rainfall.[1,2] They are found on all three of the major groups of soluble sedimentary rocks, namely limestone, gypsum and rock salt.[3,4] Recently published studies of solution flutes include Mottershead,[5,6] Vincent[7] and Ginés[8] and Crowther[9,10] in respect of limestone sites. Stenson and Ford[11] and Calaforra[12,13] have described flutes on gypsum. Mottershead *et al.* present data on flute morphology across all three groups of soluble rocks.[14]

Two substantive models have previously been advanced to explain the process of solution flute development,[15,16,17] both of which rely on hydrochemical considerations alone.

Bögli's model is based on consideration of solute concentration in long profile runoff.[18] Bögli explained solution flutes on limestone exclusively in terms of dissolution, arguing that flute development and extension continued until increasing distance of runoff flow across the rock surface caused the water to become saturated in respect of calcium carbonate. At this point solutional erosion would cease and the flute would terminate.

Glew and Ford proposed a model which extended hydrochemical explanations of rock solution to explain the form of both long and cross profiles.[19] They simulated the development of solution flutes on reconstituted gypsum, and also on rock salt, under controlled laboratory conditions. They proposed a model of flute development in which the limiting factor of erosion was the increasing depth of water flowing in the flute. A critical depth was attained, determined as 0.15 mm, at which impacting raindrops bringing fresh solvent were deemed unable to penetrate the flow, thus inhibiting solution of the rock surface, and thereby causing the flutes to terminate.

Recently, however, Fiol *et al.* have shown that mechanical and biotic effects are also important components of the erosion and development of a fluted surface on limestone.[20] This implies that previous models of flute development, which relied on hydrochemical considerations alone, can offer no more than a partial explanation of the formation and morphology of solution flutes.

Evidence is now presented, based on experiment and observation in both field and laboratory, of the implication of mechanical and biotic mechanisms in flute formation. The consideration of similarities and differences of flutes developed on a range of soluble lithologies provides a broader perspective from which to consider processes of flute formation. The purpose of the present paper is, therefore:

- to consider recent observations, by the authors and others, of mechanical and biotic processes in solution flute formation,
- to evaluate their implications for mechanisms of solution flute formation,
- to propose a model of solution flute formation which embraces these forces.

THE WEATHERING OF EXPOSED SOLUBLE ROCKS

Evidence has recently emerged that the breakdown of soluble rocks exposed at the ground surface is a more complex process than formerly recognized in purely hydrochemical models. Observations by scanning electron microscope of rock surfaces undergoing breakdown provide detailed evidence of their response to erosional stresses. This evidence includes both detailed micro- and nano-topography, and the presence of particulate matter both on eroding surfaces and in runoff draining from them.

Chemical weathering processes tend to concentrate at loci of dissolution. These commonly include grain boundaries and weak points in crystal lattice structures. At the micro-scale, the topography of flute surfaces is shown to possess upstanding ridges, bosses or shoulders. These may exhibit rounding and smoothing which is interpreted as an effect of dissolution by the concentrated impact of the solvent on these exposed locations (Figure 14.1a). At the nano-scale, morphologies determined by the relative magnitude of these types of dissolution have been recognized on both gypsum and limestone by Forti[21] and include channels and etch pits. Such dissolution effects are spectacularly present on surfaces of some gypsum samples, which can be very intricately differentially etched. Halite surfaces generally show only shallow and subdued etching, commonly with shallow intergrain channels (Figure 14.1b).

Biotic effects, including those of lichen, algae and moss, in the surface weathering of rock have been recognized by a range of authors.[22,23,24,25,26,27,28] These authors show that colonized rocks are

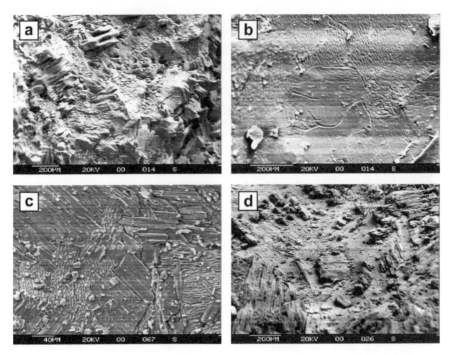

Figure 14.1 (a) Flute sidewall showing rounding of promontories by rainbeat and dissolution, Vallada, Spain, ×70. (GMFL2.3) (b) Halite nanotopography showing shallow etching along crystal boundaries and crystalline lattice, Cardona, Spain ×116. (CC5.2) (c) Crystal boundary etching in gypsum. Some loose particles released lie on the rock surface. Vallada, Spain, ×290. (VL1.5) (d) Gypsum surface with a clitter of grains detached by crystal boundary dissolution and subsequent mechanical action. Gobantes/Meliones, Spain ×83.

subject to both chemical attack by organic fluids, and mechanical dislodgement of small particulate material by fungal hyphae. Limestone flute surfaces are commonly pitted with algal borings, in the form of quasi-cylindrical holes and tunnels some 10 μm in diameter. They have the dual effects of both increasing the surface area available for dissolution reactions, and also creating a complex and fragile nanotopography, which includes isolated and unsupported horns, and other delicate lattice forms. Within the rocks in the present study, however, these effects appear to be restricted exclusively to limestone, since extensive searching by the authors by scanning electron microscope has failed to reveal any visible signs on gypsum and halite flute surfaces of the kind of biotic activity described from limestone. Furthermore, the limited literature of gypsum and salt flutes, in contrast to that for limestone, makes no mention of biotic activity.

Mechanical effects in the breakdown of these rock surfaces can be both inferred and observed. The weathering out of grain boundaries prepares individual grains for release under mechanical stress (Figure 14.1c). Many of the nanoforms created by chemical and biotic weathering processes are very fragile, and will therefore be susceptible to the mechanical forces of impacting raindrops, in particular the shear forces of drops impacting quasi vertically on the steep slopes of the flute ridges. The delicate blades and horns created by etching and algal boring are, when subsequently exposed following the decay of the organism, susceptible to brittle fracture. The presence on flute surfaces of particulate grains, commonly 10–20 μm in diameter but up to 100 μm depending on rock texture, represents material which has become mechanically detached from the mineral surface (Figure 14.1c). In some cases a litter of particulate matter is present on the flute surface (Figure 14.1d), especially in the case of gypsum with a contorted petrographic structure, indicating that mechanical detachment is likely to be a significant influence. The very subdued nanotopography of salt flute surfaces is associated with a sparsity of particulate matter.

The evidence presented above indicates that, in addition to chemical effects, biotic and mechanical processes also play a significant part in the breakdown and erosion of soluble rocks which support fluted surfaces. There are, furthermore, significant differences between the major groups of soluble rocks in the relative significance of these sets of processes.

The consequence of this range of processes operating at the point of rock breakdown can be observed in the output of mineral debris from the surface of soluble rocks. Several recent pertinent observations have been made. Zendri *et al.* subjected samples of two types of limestone to simulated acid rain in the laboratory, and observed mass loss in particulate form amounting to 1–3.5% of that in solutional form.[29] In this case the sediment is derived from dissolution of calcite cement to release fossil fragments, and the weakening of stylolites by the addition of water to release granular particles.

Initial observations by the authors of direct runoff from a gypsum surface under field conditions reveal sediment concentrations of 100–500 mgl^{-1}.[30] The sediment, of fine sand grade, comprises crystal fragments of gypsum, implying that partial crystal dissolution and dissolution along crystal boundaries alone are sufficient to release solid material. These observations, as yet preliminary, suggest that some 10–20% of the removal of the surface of this soluble rock is caused by mechanical rather than chemical processes. This implies that it is not apparently necessary for complete dissolution of the rock to occur in order to cause erosion of such a bare rock surface; limited solution alone is sufficient to permit particle detachment by mechanical forces, thereby creating a source of sediment.

Fiol *et al.* have carried out detailed laboratory simulations of rainfall on limestone surfaces.[31] They demonstrate that the rate of particulate loss from a bare rock surface amounts to 48% of the total loss, the remainder being in solution. On lichen-covered surfaces the particulate loss is lower, but still remains a major component of total loss at 35%. The particulate material lost was shown to comprise particles <5 μm in size and crystal aggregates ranging from 20–75 μm.

It follows from these observations that a full consideration of the nature of flute forming processes must include processes of mechanical erosion, and recognize the effects of biotic processes in preparing the rock surface, in addition to hydrochemical considerations. The single-process models put forward to date can offer no more than partial explanations of solution flute formation.

A MORPHOLOGICAL MODEL OF FLUTE DEVELOPMENT

A qualitative model of flute development was proposed by Mottershead which considered the geometry of both long and cross profile forms according to relative rates of lowering of flute channels and divide crests.[32] Recent field observation of flute form across different lithologies, accompanied by advances in knowledge of flute erosion processes, now permits the speculative development of this model to embrace both process and lithology. The model takes separate account of both mechanical and chemical erosion processes.

A new perspective on flute formation is provided by a consideration of mechanical stresses. For a given site and rainstorm the applied mechanical stress will tend to be uniform across the generalized plane of the fluted surface. Rainbeat impacts initially on the divides (and their slopes). The divides are the positive and therefore dominant features of the fluted slope. It follows that flute channels are subordinate forms, which develop to evacuate surplus runoff and debris from the rock surface. This perspective implies a shift of emphasis away from the flute channel and toward the divide, which may be regarded as the defining component of solution flutes, as a focus for explaining the development of fluted slopes. Crowther implicitly moves toward this position with a discussion of divide cross sections as cusps.[33]

Raindrop impact (rainbeat) creates mechanical stresses on an exposed rock surface. The mechanical force applied by each raindrop will vary according to its mass and terminal velocity. For a standard raindrop, the nature of the force applied will vary according to the angle of incidence on the fluted surface. Assuming a vertical fall, the impact will resolve predominantly into a shear force on a steep gradient, whereas on a low gradient the shear force will be subordinate to compression. Rocks are less resistant to shear forces than compressive forces, by an approximate order of magnitude.[34] Because of the lesser resistance of solid rock material to shear, there is substantially greater potential for mechanical damage, and the detachment by shearing of protruding parts of crystals or grains, when a raindrop falls on a steeply sloping surface. On a lower angle slope, where compression dominates, the greater resistance to compressive stresses means a lower potential for such mechanical damage to rock.

The potential for chemical erosion by dissolution will be a function of many factors – volume of solvent supply per unit area, aggressivity of the solvent, residence time at or near the rock surface, nanotopography, thickness of the water film, and turbulence of flow. Dissolution will be maximized when the water is at its most chemically aggressive. This will be at the point of impact with the rock, before any mineral solution has been undertaken, and will diminish with increasing contact time as the water takes up ions in solution and subsequently drains across the rock surface. Subsequent to impact, drainage of water across a smooth rock surface may be as a thin film of slow laminar flow. The nature of flow is likely to be influenced by the nanotopography of the rock surface, by which surface roughness may increase frictional resistance to flow, create depression storage and create conditions for throughflow in the case of a partially disintegrated surface. Clearly the varying nature of nanotopography of different rock types creates substantial scope for variation in surface flow conditions. Where the runoff becomes concentrated into a channel, for example, within a flute channel, then potential for turbulent flow exists. This is likely to be influenced by the magnitude of the flow, and thus by the magnitude of the flute channels. It appears possible that both film flow and turbulent flow may be penetrated by impacting raindrops, perhaps up to a threshold depth such as that postulated by Glew and Ford.[35]

It is now possible to consider the impacts of these varying mechanical and chemical processes in both cross section and long profile, and their implied variation between different lithologies.

Cross profile form

The implications of the relative magnitudes of these mechanical and chemical stresses on flute cross profiles can now be considered. It is apparent that the steep slopes of the upper flute cross section are the areas most continuously exposed to raindrop impact (Figure 14.2). Here, close to the divide, the likelihood of submergence of the surface beneath overland flow is at a minimum. It is here, where slopes are steepest on the cusp-like divides, that the aggressive effect of mechanical shear forces will have the greatest potential effect on the

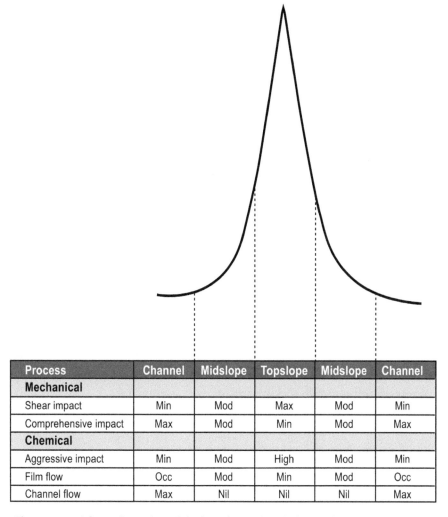

Figure 14.2 A hypothetical model of mechanical and chemical processes on the flute divide cross section profile.

intricate nanotopography. A steeply sloping surface of given length will offer a much smaller target in the horizontal plane to vertical attack by raindrop impact, thus the development of steeply sloping divide slopes may be interpreted as a response to the shearing forces of impacting raindrops. As such it may be regarded, in mechanical terms, as an equilibrium form, streamlined in response to vertical shear attack by rainbeat. The lower slope angles of the basal slope and channel floor are more subject to compressive stresses, to which the rock is substantially more resistant. Thus the semi-parabola of the divide slope profile may be interpreted as an equilibrium form, a response to the shear and compressive mechanical stresses applied by rainbeat, the dominant mechanical stress on the divides.

Chemical erosion will be effected by the impact of aggressive raindrops on the upper and midslopes with maximum exposure to fresh raindrops. It is also the case that the steeper slope increases the rock surface area exposed to solution by raindrops. By this means the volume of solvent delivered per unit of surface area is spread more thinly, bringing into direct contact with the rock surface a greater proportion of the solvent received. Downslope there is potential for an increasing layer of film flow as catchment area increases. The maximum volume of flow of surface water on the cross profile will take place along the channel floor, increasing the volume of available solvent and the theoretical potential for chemical erosion there. Turbulent flow is more likely at this location, and with it increasing potential for chemical dissolution as aggressive water is constantly renewed at the water/rock interface.

Long profile form

Flute gradients commonly lie between 30 and 70 degrees.[36] The implications of this for mechanical erosion processes are that the balance between shear and compression stresses under rainbeat will vary according to flute gradient (Table 14.1). Furthermore, Mottershead shows that flute channel long profiles are slightly concave, and it follows that the moderately diminishing downflute gradient will tend to diminish shear stress in that direction, albeit by a rather small proportion.[37]

It is likely to be of rather greater significance, however, that depth of flow increases downflute as catchment area increases, thus diminishing the probability of raindrops actually penetrating the flowing

Table 14.1 A hypothetical model of mechanical and chemical processes on the flute long profile.

Process	Upslope	Downslope
Initiation phase.		
Channel lowering : Divide lowering	>1	<1
Mechanical:		
Shear impact	Greater	Less
Compressive impact	Less	Greater
Chemical:		
Aggressive impact	Greater	Less
Film flow	Moderate	Less
Channel flow	Moderate	Greater

film and impacting directly on the rock surface, as postulated by Glew and Ford.[38] It follows from this that both mechanical impact damage and chemical erosion by impacting raindrops will tend to diminish downflute. The effects of the increasing solvent volume represented by channel flow are as yet unknown; there are potentially opposed effects on the dissolution process of turbulence, or the development of a saturated thin film of water at the channel bed. The relative influence of these flow effects at this scale are as yet unknown. There is as yet no reason to believe that Glew and Ford's interpretation of cessation (or perhaps a diminution) of chemical erosion of gypsum with the attainment of a critical depth of flow is not valid. It is also possible that an increasing depth of flow may also apply a constraint to mechanical erosion as it provides an increasing mechanical buffer against rainbeat impact to the rock surface beneath. It is equally possible that a more rapidly dissolving or more highly soluble material such as salt may behave differently in this respect. The larger flute channels characteristic of rock salt imply that greater and therefore more turbulent flow may take place there, increasing the potential for chemical erosion.[39] The nature of channel erosion is therefore likely to differ between rock materials of differing chemical and mechanical characteristics.

A flute crest tends to possess a uniform gradient through its length. Since flute crests are uniformly exposed to raindrop impact and no other significant erosive agent, and since raindrop impact may be expected to be spatially uniform across the plane of a fluted surface, the process of flute crest erosion may also be expected to be spatially uniform. A necessary corollary of this is, therefore, that flute crestlines are being lowered at the same rate both upslope and downslope.

In contrast, the flute floor has a tendency to long profile concavity. Mottershead demonstrated with respect to limestone flutes that the deepest section of the flute channel lies in the upper half of its profile.[40] Morphological, and therefore process, considerations imply that the profile can be divided into two units (Figure 14.3). It follows that flute deepening increases downslope and is greater than the crest lowering rate as far down as the point of maximum flute depth. Below the deepest section the converse is true; flute lowering downslope from this point is less than crest lowering, and as the flute shallows, the divides become redundant and are eventually

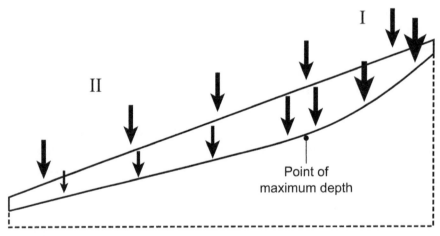

Figure 14.3 Long profile form of divide crest and flute; arrow length proportional to lowering rate during the formation phase. If the fluted slope is an equilibrium form, then it is implied that lowering rate will subsequently be uniform across the entire slope.

extinguished. It follows that the maximum erosion of the flute channel must occur at the point of maximum depth. This in turn implies that the sum of chemical dissolution and mechanical erosion within the flute channel are at a maximum at the point of inflection of the channel gradient, the point of maximum cross section, and diminish both upchannel and downchannel from this point. If Glew and Ford are correct in their argument concerning the critical depth of flow,[41] then it is here implied that this critical depth is attained at the deepest point of the flute channel.

The implications of these process considerations can now be evaluated in respect of different lithologies. Scanning electron microscope and runoff analyses have shown that mechanical and biotic processes have significant influence on the development of fluted slopes. Within the present study, observed biotic influences appear to be restricted to limestone surfaces. Mechanical erosion is shown to be effective in limestone and active in gypsum. In the case of halite, however, the available evidence suggests that mechanical processes are rather limited in effect. The implications for flute development processes are that a process model for a particular lithology should appropriately embrace both mechanical and chemical processes in relative proportion to the different resistances of the particular rock. Resistance to chemical processes is perhaps readily determined in terms of total solubility and solution rate. Mechanical strength will depend not so much on rock strength or the hardness of individual minerals present, as on the complexity of local topography, which may prove somewhat difficult to evaluate at the nano-scale.

DISCUSSION AND CONCLUSIONS

Consideration of the topographic form of fluted surfaces suggests that rainbeat may be a very significant force in their formation and that the topographic prominences, the divides and their slope profiles, may helpfully be considered as the defining landform. The cross section form of the divides creates flute channels, which are then subordinate forms along which runoff and weathered materials are evacuated.

Recent evidence suggests that 'solutional' breakdown of soluble rocks may be associated with processes of mechanical breakdown, the influence of which has not previously been fully recognized. It follows that models of fluted slope formation should recognize the respective contributions of both mechanical and biotic processes, in addition to the chemical processes of rock breakdown. The relative proportions of these will vary according to solubility characteristics of the rock minerals present, the mechanical coherence of the minerals and their interparticle bonding, and the intricacy of the nanotopography. In particular, the influence of biotic effects is demonstrated to be a significant influence on process, but appears not to be a significant influence on the resulting macro-scale surface form.

The relative significance of mechanical and chemical erosional stresses will vary across different lithologies according to the relative resistance of a particular rock to mechanical and chemical stresses, and its susceptibility to biotic attack. It is then appropriate to consider flute form in relation to these processes. Table 14.2 sets out data on flute form drawn from across the three soluble rock groups.

Table 14.2 General flute characteristic magnitudes in relation to rock groups and properties, in rank order (Mottershead, D.N., Moses, C.A. and Lucas, G.R. (2000) Lithological control of solution flute form: a comparative study. *Zeitschrift für Geomorphology N.F.* 44: 491–512).

Rock type	Solubility	Rock strength	Biotic action	Flute width	Flute depth	Flute length	Cross section area	Width/depth ratio
Limestone	3	1	Y	1	2	1	2	1
Salt	1	2	N	2	1	2	1	3
Gypsum	2	3	N	3	3	3	3	2

These data are based on 300 profiles of fluted slopes from 21 sites in Europe and Australia.[42] Flute form variables are tabulated against rock properties of mechanical strength and solubility, and the presence/absence of observed biotic action. Rank ordering of variables enables elementary comparison to be made between rock groups.

It appears from Table 14.2 that, at this general level, positive rank order correlations exist between rock strength and both flute width and flute length. Soluble rocks formed of mechanically stronger minerals support flutes which are larger in both the major dimensions. Mechanical strength thus appears to determine overall flute magnitude. Higher mechanical strength is associated with higher and longer divides, which may be interpreted as indicating that the defining cause of divides is resistance to mechanical stresses. There is also a positive rank order correlation between flute depth and cross sectional area, as would be expected on account of the geometric dependence of the latter on the former.

An inverse rank order correlation exists between solubility and width/depth ratio. This indicates that relatively deeper flutes are present on the more soluble rocks. A positive feedback exists as the flute channels concentrate the available solvent. This would appear to offer a reasonable explanation of the great relative depth of flutes in salt, the most soluble material in this study, whilst the moderate mechanical strength is capable of maintaining divide height despite the high solubility.

Despite the apparent differences between the three major groups of soluble rock in respect of the balance of mechanical, biotic and chemical processes of rock breakdown, there is nevertheless a fundamental similarity in the form of fluted slopes across the groups. This appears to emphasize an overriding commonality of process in their formation. Given that observed biotic effects are apparently restricted to limestone, it appears that this commonality of process is dominated by mechanical and chemical processes alone, and that the presence of algae and lichen does not have any fundamental influence on the process of flute formation. The presence of lichen would appear merely to limit the rate of surface erosion.[43]

Several recent researchers suggest that fluted slopes represent an equilibrium form.[44,45,46] If this is the case, then it is a major implication of equilibrium that, in order to preserve this form, the vertical lowering rate should be spatially uniform. Thus, in order for the form to be maintained, it follows that both divides and flutes are being lowered at the same rate. This condition cannot, however, exist during the phase of flute initiation and development, otherwise differentiation of an initial plane surface into divides and flutes would not occur. It therefore follows that the balance of erosion processes must differ between the initial phase of development, and the subsequent phase of maintenance. In the former, the slope becomes differentiated into flutes and divides; in the latter the morphology of fluted slope is maintained as an equilibrium form. The process considerations in this study, therefore, are concerned with field forms in the maintenance phase, in which the fluted form is already in existence. Maintenance, however, does not necessarily mean constancy. Attention has been drawn to the tendency toward long profile concavity in solution flute form, and there is no *a priori* reason why this concavity should remain constant. Indeed, analogy with other natural channel forms suggests that the flute channel profile should evolve over time. There is currently an apparent paucity of data on flute long profile form. The collection of data on long profile form and its variation is an aspect of flute morphology which could usefully inform future studies.

This paper has drawn attention to the role of mechanical processes in flute formation. Mechanical processes impose stresses in the vertical direction. Studies of the morphometry of fluted slopes, however, have commonly considered flute cross-section profiles in the slope-normal plane. It is therefore more appropriate that flute cross-section form should be assessed in the vertical, rather than the slope-normal, plane. Flute cross sections in the vertical plane are deeper by a factor equal to the reciprocal of the cosine of slope angle, thus creating longer and steeper divide sideslopes than hitherto utilized in analysis. Any future dynamic modelling of flute and divide formation will need to embrace this.

This consideration of flute forming processes indicates that in addition to chemical processes, mechanical and biotic stresses play an important role in flute formation. These processes are likely to vary in significance between flutes and divides. Rock material properties which influence resistance to mechanical and biotic stresses should be taken into account in considering flute forming mechanisms. At the rock group level, clear lithological control of flute characteristics is demonstrated, although the mechanisms through which this control is exercised are clearly complex and difficult to interpret.

Acknowledgements

Norman Jenkinson provided very helpful assistance with the scanning electron microscope. Cherith Moses commented on an earlier draft of this paper. Ann Chapman drafted the diagrams. Edge Hill College of Higher Education contributed towards field expenses.

References

1. Allen, J.R.L. (1984) *Sedimentary Structures: Their Character and Physical Basis*, Volume 2. Elsevier, Amsterdam.
2. Ford, D.C. and Lundberg, J. (1987) A review of dissolutional rills in limestone and other soluble rocks. *Catena Supplement* 8: 119–140.
3. White, W.B. (1988) *Geomorphology And Hydrology Of Karst Terrains*. Oxford University Press, Oxford.
4. Ford, D.C. and Williams, P. (1989) *Karst geomorphology and hydrology*. Unwin Hyman, London.
5. Mottershead, D.N. (1996) A study of solution flutes (Rillenkarren) at Lluc, Mallorca. *Zeitschrift für Geomorphology N.F. Supplementband* 103: 215–241.
6. Mottershead, D.N. (1996) Some morphological properties of solution flutes (Rillenkarren) at Lluc, Mallorca. In Fornós, J.-J. and Ginés, À. (Eds) *Karren Landforms* Palma: IUCN/IGU: 225–238.
7. Vincent, P.J. (1996.) Rillenkarren in the British Isles. *Zeitschrift für Geomorphology N.F.* 40: 487–497.

8   Ginés, A. (1996) Quantitative data as a base for the morphometrical definition of rillenkarren features found on limestones. In Fornós, J.-J. and Ginés, À. (Eds) *Karren Landforms* Palma: IUCN/IGU: 177–191.
9   Crowther, J. (1997) Surface roughness and the evolution of karren forms at Lluc, Serra de Tramuntana, Mallorca. *Zeitschrift für Geomorphology N.F.* 41: 393–407.
10 Crowther, J. (1998) New methodologies for investigating Rillenkarren cross-sections: a case study at Lluc, Mallorca. *Earth Surface Processes and Landforms* 23: 333–344.
11 Stenson, R.E. and Ford, D.C. (1993) Rillenkarren on Gypsum in Nova Scotia. *Géographie Physique et Quaternaire* 47: 239–243.
12 Calaforra, J.M. (1996) Some examples of gypsum karren. In Fornós, J.-J. and Ginés, À. (Eds.) *Karren Landforms*, Palma: IUCN/IGU: 253–260.
13 Calaforra, J.M. (1998) *Karstologia de yesos*. Almeria: Universidad de Almeria: 384.
14 Mottershead, D.N., Moses, C.A. and Lucas, G.R. (2000) Lithological control of solution flute form: a comparative study. *Zeitschrift für Geomorphology N.F.* 44: 491–512.
15 Bögli, A. (1960). Kalklösung und Karrenbildung. *Zeitschrift für Geomorphology Supplementband* 2: 4–21.
16 Bögli, A. (1980) *Karst hydrology and physical speleology*. Springer-Verlag, Berlin.
17 Glew, J.R. and Ford, D.C. (1980) A simulation study of the development of Rillenkarren. *Earth Surface Processes* 5: 25–36.
18 Bögli, *op. cit.* (1960)
19 Glew and Ford, *op. cit.* (1980)
20 Fiol, Ll., Fornós, J.-J. and Ginés, À. (1996) Effects of biokarstic processes on the development of solutional rillenkarren in limestone rocks. *Earth Surface Processes and Landforms* 21: 447–452.
21 Forti, P. (1996) Erosion rate, crystal size and exokarst microforms. In Fornós, J.-J. and Ginés, À. (Eds) *Karren Landforms* Palma: IUCN/IGU: 261–276.
22 Viles. H.A. (1984) Biokarst: review and prospect. *Progress in Physical Geography* 7: 523–542.
23 Viles, H.A. (1987) A quantitative scanning electron microscope study of evidence for lichen weathering of limestone, Mendip Hills, Somerset. *Earth Surface Processes and Landforms* 12: 467–473.
24 Cooks, J. and Otto, E. (1990) The weathering effects of the lichen Lecidea Aff. Sarcogynoides (Koerb) on Magaliesberg Quartzite. *Earth Surface Processes and Landforms* 15: 491–500.
25 Moses, C.A., Spate, A.P., Smith, D.I. and Greenaway, M.A. (1995) Limestone weathering in Eastern Australia. Part 2: Surface micromorphology study *Earth Surface Processes and Landforms* 20: 501–514.
26 Moses, C.A., and Viles, H.A. (1996) Nanoscale morphologies and their role in the development of karren. In Fornós, J.-J. and Ginés, À. (Eds) *Karren Landforms* Palma: IUCN/IGU: 85–96.

27 Forti, *op. cit.* (1996)
28 Fiol *et al.*, *op. cit.* (1996)
29 Zendri, E., Biscontin, G., Bakolas, A. and Finotto, G. (1996) Simulated study on the chemical and physical decay of the acid rain on carbonate stone. In Riederer, J. (ed.) *Proceedings of the 8th International Congress on the Deterioration and Conservation of Stone*, Berlin, Germany, 30 September–4 October, 1996: 273–279.
30 Mottershead, D. and Lucas, G. (2001) Field testing of Glew and Ford's model of solution flute evolution. *Earth Surface Processes and Landforms* 26: 839–846.
31 Fiol *et al.*, *op. cit.* (1996)
32 Mottershead, D.N. (1996) Some morphological properties of solution flutes (Rillenkarren) at Lluc, Mallorca. In Fornós, J.-J. and Ginés, À. (Eds) *Karren Landforms* Palma: IUCN/IGU: 225–238.
33 Crowther, *op. cit.* (1998)
34 Billings, M.P. (1954) *Structural Geology*. Prentice Hall, Englewood Cliffs, N.J.
35 Glew and Ford, *op. cit.* (1980)
36 Ford and Williams, *op. cit.* (1989)
37 Mottershead, D.N. (1996) Some morphological properties of solution flutes (Rillenkarren) at Lluc, Mallorca. In Fornós, J.-J. and Ginés, À. (Eds) *Karren Landforms* Palma: IUCN/IGU: 225–238.
38 Glew and Ford, *op. cit.* (1980)
39 Mottershead *et al.*, *op. cit.* (2000)
40 Mottershead, D.N. (1996) Some morphological properties of solution flutes (Rillenkarren) at Lluc, Mallorca. In Fornós, J.-J. and Ginés, À. (Eds) *Karren Landforms* Palma: IUCN/IGU: 225–238.
41 Glew and Ford, *op. cit.* (1980)
42 Mottershead *et al.*, *op. cit.* (2000)
43 Fiol *et al.*, *op. cit.* (1996)
44 Glew and Ford, *op. cit.* (1980)
45 Mottershead, D.N. (1996) A study of solution flutes (Rillenkarren) at Lluc, Mallorca. *Zeitschrift für Geomorphology N.F. Supplementband* 103: 215–241.
46 Crowther, *op. cit.* (1998)

CONTACT ADDRESS
Department of Geography,
University of Portsmouth,
Buckingham Building,
Lion Terrace,
Portsmouth, PO1 3HE
United Kingdom

15 Weathering Scales, Landscapes and Change: Some Thoughts on Links

R.J. INKPEN

ABSTRACT

Weathering is defined by relations rather than being a fixed phenomenon. The observer is part of these relationships and has a key role to play in defining the entities of interest and the scale of study. Form identification is part of the dialogue the observer undertakes with reality. The forms and processes identified cannot be treated separately from the measurement systems used to identify and monitor them. Within such a framework explanation cannot be reduced to process alone, but should be seen as developing out of the interrelationships between process, form, material and environment. The unique combinations of these factors each provide constraints on the explanations that are viewed as plausible.

INTRODUCTION

Weathering studies have tended to focus on small-scale forms and process based studies. Whilst these studies provide insights into the potential processes responsible for weathering, understanding of relationships between processes and weathering forms on buildings, and at larger scales, is relatively poorly developed. This paper suggests a number of ideas that may help link weathering across spatial and temporal scales. Initial discussion focuses on the relational nature of weathering and the central role of the observer. The relationships between reality and measurement systems used to identify and quantify weathering are then explored. Finally, some consideration is given to hierarchical analysis of scales and the potential of the landscape metaphor in understanding changes in weathering forms. Although most of these ideas have been discussed elsewhere, they have not generally been articulated and discussed within the context of weathering. It is hoped that articulation of these ideas will provide at least some food for thought and at best new perspectives on some old problems.

Weathering is usually expressed in studies by reference to discrete spatial and temporal scales, involves some reference to alteration *in situ*[1] and implies that weathering operates within a predefined area or volume. However, as noted by systems analysis pioneers in geomorphology, Chorley and Kennedy,[2] the observer is central to defining the entity of interest. The observer defines the system, the variables inside its boundaries and the externalities outside. Whilst not denying that this approach is vital for isolating the phenomenon of study, it inevitably highlights weathering as something that happens at a single observer defined scale. However, weathering cannot be studied by isolating it from the relationships at other scales that, at least partially, define it. Trudgill,[3] for example, shows that dissolution of limestone can be thought of as a five stage process of reactant delivery, reactant absorption onto the surface, reaction with the surface, sorption of products and removal of products. Each stage, although divisible from the other, is interdependent and explicable, only in relation to the other stages. The whole sequence depends upon the supply of reactants and their removal. Trudgill identifies dissolution as the essential framework for the process, but not the contingent relationships that determine whether it will happen and at what rate.[4] The identification of the essential relationships that define dissolution depends however, upon abstraction and identification by the observer. Dissolution is not a given, the observer has to have a model of expected behaviour of a substance, an acid, with another substance, a base, to build up the essential framework for the reaction stages. This model is then hunted for in reality. Its identification as a real weathering phenomenon by the observer may require isolation of an appropriate section of reality at an appropriate scale. Similarly, it may require the use of an appropriate measurement technique, such as runoff sampling, to identify its occurrence.

The above illustration highlights that the observer is not a passive agent in the identification of weathering. The observer defines the entities of interest and the manner in which they are expected to operate. Based on these expectations, measurements are made that are expected to identify the phenomenon believed to exist, often in a manner that ensures that only the predicted model is measured. Indeed, it could be argued that understanding of the phenomenon of weathering is only possible through these models and that weathering does not exist outside of the models we use to identify it. This approach matches the critical realist frameworks of Bhaskar[5,6] and Collier.[7] By incorporating the observer as a key part of the study, the transient and relational nature of the models used to understand an intransient, but unknowable, reality, become clearer.[8] Baker similarly highlights the role of the observer as an interpreter of signs rather than as a direct mirror for reality within the sign-object-interpretant triumvirate.[9] Baker,[10] along with Simpson[11] and Frodeman,[12] view geomorphology as a dialogue or interrogation between the observer and a reality that is only knowable via the mental constructs of the observer. It is the interplay between the observer and reality that creates the entities, the phenomenon of study and reification.

Classification schemes for weathering forms illustrate the role of this interplay in defining reality. Although there is a range of schemes, there is no agreed set of definitions for different weathering forms. There is an assumption that schemes are compatible in some way as each refers to the same reality. Within this view a highly complex and differentiated scheme such as the Fitzner classification[13,14] would be viewed as a more complex version of a simpler, less differentiated scheme such as Martin et al.[15] The two schemes have at their base an attempt to understand the array of weathering forms observed on buildings, and each classification, by its very nature, tries to include all forms. Both are also concerned with identifying forms for purposes of conservation and use visual criteria at the scale of a single wall. However, even though both interrogate the same reality with similar initial transient models, they diverge in their precise allocations. Fitzner's classification identifies forms according to a set of hierarchical criteria and with a number of assumptions concerning processes of formation.[16] Martin et al only retain a few process assumptions,[17] and approach the classification task with only visual criteria in mind and no hierarchy of forms. The two schemes are not, therefore, simply more or less complicated versions of reality, they are different in their detailed vision of what weathering is and can produce.

WEATHERING AND MEASUREMENT

If the context and transient models employed in weathering studies are consistent then there will be general agreement, in the definition of the entities, methods of study and the results obtained. Rhoads and Thorn suggest that such an outcome results from the use of the theoretical model approach.[18] Within this approach objectivity, which assumes real properties of entities existing independent of measurement, cannot be maintained. Murdoch claims that properties of entities are instead relational.[19] The act of measurement produces observations that express the relations between the system under study and the apparatus used to measure the system. An indivisible whole is formed between the measurement system and the phenomenon under study. The measured phenomenon only possesses the attribute given to it by the measurement system because of that measurement system. Altering the measurement system alters the relations between its elements and so changes the observations. Alternatively, if a consistent measurement system is maintained, which includes what is thought to exist and capable of measurement, and a consistent and convergent set of values can be obtained for a phenomenon. This does not mean, however, that convergence of values gives an increased reality to the phenomenon. Different instruments may be used, but they can still be used within a common measurement system with respect to the transient model of reality maintained by the observer. In contrast to the idealist stance of Harrison and Dunham,[20] a relational view still retains an external reality with which the observer has a dialogue. This may only be knowable via transient models, but these are formed through, and therefore informed by, the intransient reality in the dialogue.

The role of the mental constructs of the observer can also be seen in a study of disrupted surfaces on the School of Geography wall in Oxford in Inkpen et al.[21] In this study a single form, surface disruption, is mapped using two different units of study: single blocks and an entire wall. Images of the wall were transferred into a Geographical Information System and surface disruption was mapped (on the basis of visual criteria such as colour, texture and morphology) as a percentage cover on each block and as percentage cover of the whole wall. Mapping the wall was not based on assumptions about the location or confinement of forms. Each method resulted in similar maps as each interrogated reality in a similar manner. Visually, the latter method seemed to match the observer's impression of the distribution of surface disruption. Both distributions have no existence independent of the measurement system within which they are contained.

Hall's recent work on freeze-thaw weathering also highlights how a measurement system can define an entity.[22,23,24] Hall noted that the perceived dominance of freeze-thaw as the agent of weathering in cold environments is based on a circular argument. Products of freeze-thaw are angular so any angular debris in a current or former cold region may be assumed to have resulted from freeze-thaw. The importance of freeze-thaw activity tends to be backed up by measurements of air temperature at resolutions of 15 minutes to 1 hour or longer. Properties such as the intensity of freezing or the cycling of temperature across 0°C are then taken to be indicative of potentially important parameters for weathering. Through this, freeze-thaw becomes an entity, a distinct process capable of manipulation, which both interacts with rock and defines the rock properties of importance in the interaction.

In this series of papers, Hall suggested that although it is easier to measure, air temperature is in reality a poor surrogate for rock surface temperature. Similarly, Hall suggested that the timescales used in measuring temperature are too long for effects other than freeze-thaw to be defined as of importance. Thermal stress may also be a significant mechanism in regions normally identified as dominated by freeze-thaw. Hall demonstrated that to identify thermal stress, temperature changes in rock should be recorded at a resolution of at least 1 minute, preferably 30 seconds, as it is changes in temperature gradient within the rock that generate thermal stress. Both freeze-thaw and thermal stress are therefore defined by the measurement system and have no existence outside of the system. Hence, they both represent a specific outcome of interrogating reality in a particular way.

WEATHERING AND SCALE

Schumm and Lichty, in their classic paper on scale,[25] viewed differences in scale as determining which variables are dependent and independent in any analysis of landscape change. However, 'real' explanation increasingly involves recourse to physical and chemical processes responsible for change – a so-called reductionist view. Explanation is equated with establishing the rules, the laws, which govern behaviour, even at the integrated level of larger scales. In a reductionist view, forms identified at larger scales are perceived as reducible to fundamental physical and chemical processes. If, for example, an agent such as salt is identified, the weathering mechanism becomes clear. The forms produced reflect the action of the agent and their characteristics are reducible to the mechanisms of salt expansion such as crystallization, hydration or thermal expansion. The physics and chemistry of salt weathering reflect the potential action of salt, alone they do not explain the weathering produced. As Smith noted,[26] weathering is a product of the inter-related factors of material, environment, form and process. In the case of salt weathering, salt crystal growth would not be possible without a supply of saturated solution or water or heat depending on the growth mechanism. Salt expansion would cause no alteration if stress was not applied that caused strain beyond the elastic limit of that specific volume of rock. It is pointless to talk of salt weathering in the abstract as salt weathering requires a relationship between salt and rock. The definition of salt weathering requires reference to both components.

The above scenario, in which the factors of rock properties and environment are considered, is sometimes referred to as the configuration or context within which the processes operate.[27,28] Configuration and context are not generally given the same prominence in explanation as 'process', which is viewed as general and dynamic rather than specific and passive. Process based explanations may identify general weathering relationships under strictly controlled conditions, such as in laboratory-based salt weathering experiments, but they do not provide complete explanation. A more appropriate view of explanation and its changing nature with scale may be provided by hierarchy theory.[29,30,31] Within this theory, explanation cannot be offered at any scale of enquiry independently of the levels immediately above and below it. Explanation can be reduced, but to a consideration of these three levels only. Changes beyond these levels may occur, but they are always mediated through the three levels and so are viewed as affecting their functioning rather than as causes in their own right. Each level is referred to as a 'holon' and they are linked by an exchange of information, matter and energy. Each level has a set of fixed properties and relationships that determine its structure and functioning. Within these constraints there are a range of potential strategies that a holon can follow. The relationships between the three levels determine the strategy followed. Following a particular strategy at one level influences the other levels and the strategies followed there. In this manner, seemingly unconnected levels can interact to a degree dependent upon the strength and nature of exchanges. Importantly, the exchanges and their impact upon strategies cannot be generalized either spatially or temporally. The individual history and structure of relationships provides the basis for strategy development and for explanation.

Understanding a natural system therefore requires identification and understanding of processes at the appropriate scale and the dynamic context of the system itself. Within biology such a dynamic view has been provided by the landscape metaphor.[32,33,34,35] Grossly simplified and without the many complications and simplifications, this view suggests that organisms have the capability to explore a static or changing fitness landscape, but are constrained in their wanderings by their current structure and potential genetic changes as to the rate and direction of exploration. All scales, the genetic, the historical and environmental are focused on the individual organism and contribute to an explanation of its development. Within weathering studies a similar claim might be made for the weathering form. Although this view could be seen as a restatement of the content/context or configurational views, it does not admit to a distinction between content and context. The form, is simultaneously content and context. The weathering form becomes the focus of the organizational/compositional relationships within the system[36] and also the agent of altering and maintaining these relationships. A very stable form will both reflect and maintain the structural and organizational relationships that produced and are producing it. A very dynamic form or deforming structural relationships will open up a range of possible pathways for change. All change is constrained by the fitness landscape within which the form lies as well as change in the form itself which alters that fitness landscape.[37]

A good example of the constrained nature of explanation in weathering is provided by Smith et al.[38] and Smith.[39] In these papers a conceptual model was developed for surface loss on non-calcareous sandstone. Surface weathering forms are viewed as moving through stages in a set of preconditioned sequences. The final weathering form for many of the sequences is tafoni, but there are a number of stages or pathways that any individual surface could follow to reach this end stage. Tafoni is not the only end form, however, as breaching of a black crust can also represent the end of the sequence, but to reach this the surface has to traverse the whole set of stages. The conceptual model could be thought of as representing the potential pathways that a non-calcareous stone surface in a polluted environment could move along. Each pathway is itself constrained by the set of relationships that define the interactions between the stone, weathering processes, the environmental conditions and the past state of the surface. Physical and chemical processes can therefore limit which types of changes are possible, but they do not define the precise nature of each pathway nor the actual sequence of change undergone. Pathways are instead defined by the interaction of factors of which processes are only a part. Movement along one pathway constrains the potential future forms that can develop on the surface. In this manner current form can influence the nature of the fitness landscape and the potentiality of different pathways.

Warke and Smith also developed a model in which process constrains, rather than explains, change, through a study of two clay-rich Scrabo sandstone blocks from a building in Belfast.[40] Analysis of the salt concentrations and distributions in both blocks showed high gypsum concentrations in the outer 20–30 mm of the stone. Block A had high concentrations down to about 60 mm, but exhibited little sign of deterioration. Block B, on the other hand, showed extensive signs of deterioration including flaking characteristic of this stone type. Warke and Smith suggested that block A represents the preparatory phase in salt weathering with significant subsurface alteration occurring, but no outward signs of deterioration; block B was regarded as having crossed the stress/strength threshold for decay where positive feedback has been initiated within the stone.[41] The positive feedback leads to continued decay and surface retreat through the migration of a salt-enriched zone back into the stone substrate. Block A represents a potential for deterioration, and may progress along a 'pathway' of deterioration as internal or external thresholds become breached.

In both examples considered above, development of conceptual models of deterioration relies on the observer having an understanding of the physical and chemical basis of decay, and an understanding of process. These constrain the types of changes that can occur, but whether such changes occur or not depends upon the interactions of other factors. In both cases, constraints of process, material, environment and form provide the framework for developing a plausible narrative for change. The narrative has to conform to the dialogue being conducted between the observer with reality. The narrative can alter as the dialogue changes either though different models or different measurement systems. It is thus clear that the observer becomes vital in defining the entities of interest and the nature of the plausible narrative used to explain them. This might suggest, however, that the observer also defines the scale of study.

Schumm and Lichty's framework views scale as a rigid framework rooted in absolute space and time.[42] Change is something that occurs within the solid, pre-defined framework. It is the observer who defines the entities of study via a dialogue with reality, so it could be argued that the entities themselves define the scale. This implies that the appropriate scale of study could alter as the entities undergo change. The continuation of an entity of interest in a form identifiable by the observer provides the basis for the appropriate scale of study. This more fluid view of spacetime is reflected in recent work on object-oriented GIS. Raper and Livingstone,[43] for example, illustrate that modelling geomorphological entities may more appropriately use a four dimensional relative space than a static absolute space. In this framework time becomes a property of the entities, so the entities define the appropriate scale of study. Putting scale first, such as micro, meso or macro, defines the entities and so limits what is and is not viewed as real. Within this view weathering entities identified by the observer define the scale at which explanation or plausible narratives should be constructed.

CONCLUSIONS

Weathering studies have tended to focus on process based explanations. Context and relationships that define the entities of interest and hence the scale of study and explanation tend to be ignored. Explanation is located in the relations defining the form including the observer and measurement system. Focusing on the form rather than process highlights the importance of constraining elements of the relationships defining the form. These restrict the type of explanation or plausible narrative possible as well as the potential pathways of change open to a form. Within such a framework scale becomes something defined by the entities studied and not a static, fixed framework.

Acknowledgements

The author would like to thank Peter Collier, Dominc Fontana and Dave Petley for their useful discussions of some of the points in this paper at various times in its gestation. The author would also like to thank the comments of two anonymous reviewers.

References

1. Bland, W. and Rolls, D. (1999) *Weathering: An Introduction to the Scientific Principles*. Arnold, London.
2. Chorley, R.J. and Kennedy, B.A. (1969) *Physical Geography, A Systems Approach*. Prentice Hall, London.
3. Trudgill, S.T. (1985) *Limestone Geomorphology*. Longmans, London.
4. Ibid.
5. Bhaskar, R. (1986) *Scientific realism and human emancipation*. Verso, London.
6. Bhaskar, R. (1989) *Reclaiming Reality: A Critical Introduction to Contemporary Philosophy*. Verso, London.
7. Collier, A. (1994) *Critical Realism: An Introduction to Roy Bhaskar's Philosophy*. Verso, London.
8. Rhoads, B.L. and Thorn, C.E. (1994) Contemporary philosophical perspectives on physical geography with emphasis on geomorphology. *Geographical Review* 84: 90–101.
9. Baker, V.R. (1999) Geosemiosis. *Geological Society of America Bulletin* 111:633–645.
10. Ibid.
11. Simpson, G.G. (1963) Historical science. In Albritton, C.C. Jr. (ed.) *The Fabric of Geology*. Addison-Wesley, Reading, Pennsylvania: 24–48.
12. Frodeman, R.L. (1998) Geological reasoning: Geology as an interpretative and historical science. *Geological Society of America Bulletin* 107: 960–968.
13. Fitzner, B., Heinrichs, K. and Kownatzki, R. (1992) Classification and mapping of weathering forms. *Proceedings of the Seventh International Congress on Deterioration and Conservation of Stone*, Lisbon, 15–18 October, 1992. Laboratorio Nacional de Engenharia Civil: 957–968.
14. Fitzner, B., Heinrichs, K. and Volker, M. (1996) Monument mapping – a contribution to monument preservation. In Zezza, F. (scientific editor) *Origins, Mechanisms and Effects of Salt on Degradation of Monuments in Marine and Continental Environments*, Proceedings European Commission Research Workshop, March 25–27, 1996, Bari, Italy, Protection and Conservation of the European Cultural Heritage Research Report No.4: 345–355.
15. Martin, B., Mason, D. and Bryan, P. (2000) Decay mapping of polishable limestone. In *Proceedings of the Ninth International Congress on Deterioration and Conservation of Stone*, Venice 19–24 June 2000: 101–109.
16. Fitzner, *op. cit.* (1992)
17. Martin *et al, op. cit.* (2000)
18. Rhoads, B.L. and Thorn, C.E. (1996) Toward a philosophy of geomorphology. In Rhoads, B.L. and Thorn, C.E. (eds.) *The scientific nature of geomorphology*. John Wiley and Sons Ltd., Chichester: 115–143.

19 Murdoch, D. (1987) *Niels Bohr's Philosophy of Physics*. Cambridge University Press, Cambridge.
20 Harrison, S. and Dunham, P. (1999) Decoherence, quantum theory and their implications for the philosophy of geomorphology. *Transactions of the Institute of British Geographers* 23: 501–514.
21 Inkpen, R.J., Collier, P. and Fontana, D. (2001) Mapping decay: integrating scales of weathering within a GIS. *Earth Surface Processes and Landforms* 26: 885–900.
22 Hall, K. (1993) Rock moisture data from Livingston Island (Maritime Antarctic) and implications for weathering studies. *Permafrost and Periglacial Processes* 4: 245–253.
23 Hall, K. (1995) Freeze-thaw weathering: The cold region 'Panacea'. *Polar Geography and Geology* 19: 79–87.
24 Hall, K. (1999) The role of thermal stress fatigue in the breakdown of rock in cold regions. *Geomorphology* 31: 47–63.
25 Schumm, S.A. and Lichty, R.W. (1965) time, space and causality in geomorphology. *American Journal of Science* 263: 110–119.
26 Smith, B.J. (1996) Scale problems in the interpretation of urban stone decay. In Smith, B.J. and Warke, P.A. (eds.) *Processes of Urban Stone Decay*. Donhead, London: 3–18.
27 Simpson, *op. cit.* (1963)
28 Lane, S.N. and Richards, K.S. (1997) Linking river channel form and process: time, space and causality revisited. *Earth Surface Processes and Landforms* 22: 249–260.
29 Allen, T.F.H. and Star, T.B. (1982) *Hierarchy: Perspective for Ecological Complexity*. University of Chicago Press, Chicago.
30 Haigh, M.J. (1987) The holon: hierarchy theory and landscape research. In Ahnert, F. (ed.) *Geomorphological Models: Theoretical and Empirical Aspects*, Catena Supplement 10: 181–192.
31 de Boer, D.H. (1992) Hierarchies and spatial scale in process geomorphology: a review. *Geomorphology* 4: 303–318.
32 Wright, S. (1932) The role of mutation, inbreeding, crossbreeding, and selection in evolution. *Proceedings of the 6th International Congress on Genetics* 1: 355–366.
33 Waddington, C.H. (1940) *Organisers and Genes*. Cambridge University Press, Cambridge.
34 Thomas, R.D.K. and Reif, W-E. (1993) The skeleton space: A finite set of organic designs. *Evolution* 47: 341–360.
35 Niklas, K.J. (1999) Evolutionary walks through a land plant morphospace. *Journal of Experimental Botany* 50: 39–52.
36 Spedding, N. (1997) On growth and form in geomorphology. *Earth Surface Processes and Landforms* 22: 261–265.
37 Inkpen, R.J. and Petley, D.N. (2001) Fitness spaces and their potential for visualising change in the physical landscape. *Area* 33: 242–251.

38 Smith, B.J., Magee, R.W. and Whalley, W.B. (1994) Breakdown patterns of quartz sandstone in a polluted urban environment. In Robinson, D.A. and Williams, R.B.G. (eds.) *Rock Weathering and Landform Evolution*. John Wiley and Sons, Chichester: 131–150.
39 Smith, *op. cit.* (1996).
40 Warke, P.A. and Smith B.J. (2000) Salt distribution in clay-rich weathered sandstone. *Earth Surface Processes and Landforms* 25: 1333–1342.
41 Ibid.
42 Schumm and Lichtys, *op. cit.* (1965)
43 Raper, J. and Livingstone, D. (1995) Development of a geomorphological spatial model using object-oriented design. *International Journal of Geographical Information Systems* 9: 359–383.

For Product Safety Concerns and Information please contact
our EU representative GPSR@taylorandfrancis.com Taylor & Francis
Verlag GmbH, Kaufingerstraße 24, 80331 München, Germany